MW00635203

Linux Robotics

Programming Smarter Robots

D. Jay Newman

McGraw-Hill

New York Chicago San Francisco Lisbon London Madrid
Mexico City Milan New Delhi San Juan Seoul
Singapore Sydney Toronto

The McGraw-Hill Companies

Cataloging-in-Publication Data is on file with the Library of Congress.

1 2 3 4 5 6 7 8 9 0 DOC/DOC 0 1 0 9 8 7 6 5

ISBN 0-07-144484-X

The sponsoring editor for this book was Judy Bass and the production supervisor was Pamela A. Pelton. It was set in New Century Schoolbook by Patricia Wallenburg. The art director for the cover was Anthony Landi.

Printed and bound by RR Donnelley.

This book was printed on recycled, acid-free paper containing a minimum of 50% recycled, de-inked fiber.

To my beloved wife, Lee, who believed in me even when I didn't.

Contents

Introduction

It is traditional for an author to explain how he has come to release a book to the unsuspecting public. And who am I to defy such a tradition?

I've been a programmer for many years. I worked for Penn State and took an occasional consulting job. One such job resulted in my having a few hundred extra bucks, and so I treated myself to a toy: the Lego Mindstorms set. It really did start out as a computer case modification and then got out of hand.

Yes, the robot bug bit me. So did my dog, but that's another story. I built and rebuilt many Lego robots until I wanted to build something bigger. So I sold my Legos (which I had built up to a large collection by that time) and started building bigger robots. Eventually I started building a robot with a Mini-ITX motherboard running Linux.

Today, there are several open projects that build medium-sized robots with Linux motherboards, but when I started there wasn't much out there. I looked into the bases available to me and built one from a large R/C tank and a second from a 16-inch-diameter Zagros Robotics base. Now I have enough experience to build a robot without a kit, but back then my mechanical skills were (to be polite) lacking.

I created a Java robotics framework just to be able to use simple behavioral programming on my robots. With these tool plus many of the open source packages available on Linux, I was able to create some very interesting behavior.

I still would suggest that nobody start with a Linux robot. If you've never built a robot before, I suggest that you build a small kit robot first to see if you want to spend the time and money for a Linux robot. However, this is your choice.

I learned a lot of things while building smaller robots that I wouldn't have learned if I had jumped into building bigger ones.

And I always like to remember that it all started with a set of Legos…

I would also like to thank my editor, Judy Bass, who prodded me when I needed prodding.

Linux Robotics

Chapter

1

Starting Out

This chapter is similar to the first chapter in most robotics books. I will explain the basics of the robots I'll be using as examples, the reasons for my choices, and some other fairly basic stuff. I am assuming that you already have gone through some basic robotics books and constructed a robot or two. If you haven't, don't worry: I'll try to make this as painless as possible. It's easy to make a robot.

You need to have a robotic base to work through the examples. Because the Linux motherboards I'll be using for the examples require more power than the standard robot does, you'll need a bigger base. I would suggest a base at least 12 inches in diameter with a payload capacity of at least 30 pounds. There are several ways to get such a base: buying a kit, converting a large toy, and building your own. What you do is up to you and your skills. If you've never built a robot before, I strongly suggest buying a kit for the base.

I think you'll get much more out of this book if you construct a robot yourself and try out the examples. You can learn more from actually building things than you can from reading alone. In addition, you might find any mistakes I've made so that I can correct them for the next edition.

After building a robot, I will talk about the code needed to control it.

It might be good to read this chapter even if you have built several robots already. If you are using your own design, you still need to know how it is different from the one in the examples.

The equipment for the example robot (named Groucho) includes the following:

- iBase 890c Pentium M Motherboard running Linux (I paid enough for the robot; why should I pay for an operating system when a better one is available free?) with a Pentium M 755 (2.0 GHz) and 1 GB SDRAM. After you read this book, this may seem like nothing, but for now it is a cutting-edge motherboard. I chose this processor because its high speed should allow processor-intensive activities such as audio and video.
- A 40-GB notebook hard drive. A notebook drive is used for lower power consumption and because it has the ability to withstand more shaking than a standard drive.
- A webcam. The one I chose was a Logitech Orbit, mainly because it looked cool.
- A microphone. I have an inexpensive off-brand PC microphone that is used for voice recognition.
- A set of nonpowered speakers. They cost less than $10 and are designed for portable CD players. They don't have much volume, but they do work.
- A small LCD TFT monitor for debugging and eventually showing a face to go along with the voice.
- Twelve Devantech SRF08 sonar range finders (sonar units). These are nice because each one contains a controller that handles the actual range finding. The computer communicates with the sonar units via a serial protocol called I2C.
- One Devantech TPA81 thermopile array. This is used to detect temperature via an 8-pixel array with a field of view of 100 degrees.
- A Sharp IR Ranger. This is used to detect objects close to the front caster so that Groucho doesn't get stuck on dog toys.
- Two leaf switches. These are used as whisker sensors to detect objects close to the bottom base.
- A BDMicro MAVRIC-IIB. This is used to control the Devantech sensors.
- A Phidget Interface Kit 8/8/8. This is a USB device that is used to read the Sharp IR Ranger and the switches. It also has eight digital outputs that I'm not currently using.

- Two BDMicro REACTOR RX50 H-Bridge prototype boards (RX50s). These are another of Brian Dean's creations.
- A Linksys 802.11g WiFi PCI card. I like being able to communicate with Groucho over the Internet.
- A 16-inch-diameter base from Zagros Robotics. This base has motors with built-in encoders. I also have three additional decks on the robot. This kit has a cost of around $300 to $400. The motors are strong, but the speed is slow because Groucho weighs over 40 pounds (Figure 1-1). I like the speed because it means that running into walls doesn't break anything, though I can see the power of the motors because I have accidentally moved chairs and small children.
- One 12-volt 12-amp-hour sealed lead-acid (SLA) battery for powering the motor and two 12-volt 7-amp-hour SLA batteries for powering the electronics (they are connected in series for 24 volts and reduced to 12 volts through a switching regulator; the object is to extend the battery life).

I was going to have a second example, but I salvaged parts from my older robots to make Groucho (see Figure 1-2). This was done mainly because Stuart made too much noise for my wife. Groucho is extremely quiet.

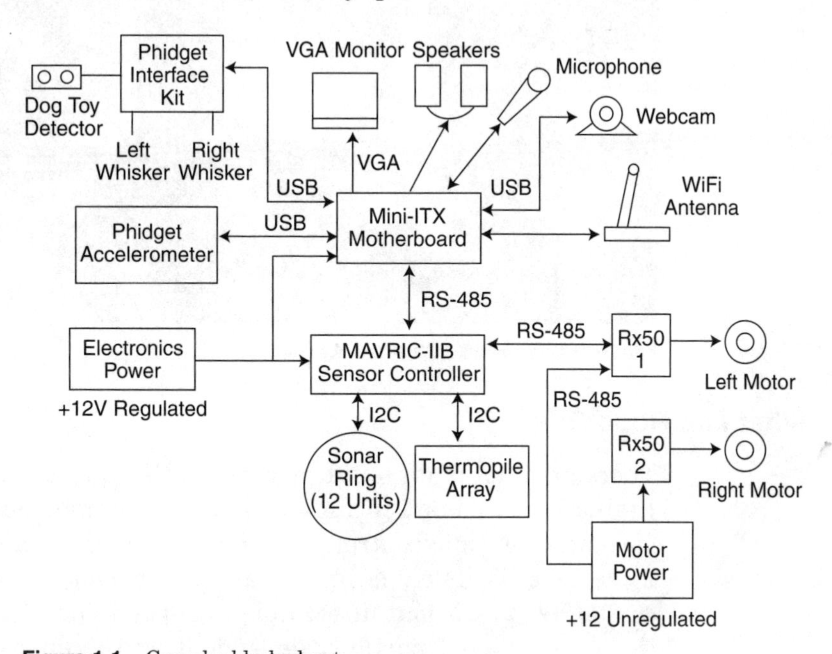

Figure 1-1 Groucho blocked out.

Figure 1-2 Groucho completed.

What Is a Robot?

There are a lot of different answers to this question, but when I talk about robots, I mean a mobile and autonomous machine with an electronic controller. *Mobile* means that the robot can move about in its environment, and *autonomous* means that it can do its job without direct human supervision. These definitions are at least partially dependent on the environment: A living room floor requires a different type of mobility, mechanics, and controller than a jungle does.

> **Note:** Not everybody uses the word *mobile* in his or her definition. For example, a simple robotic arm might not be mobile, but it could be useful (I'm thinking about using it as a backscratcher). To be honest, I'm thinking of making a charging station for my robot that will have some autonomous behavior to make sure that it's charging a robot and not the dog.

The behaviors of an autonomous robot can be very simple (go until you bump into a wall, then back up a couple of inches, turn in a random direction, and then go again—I call these robots wallbangers) or very complex (find the bride and groom at a wedding and take good wedding pictures). For this book, we'll start with wallbangers and work our way up a bit.

I do not consider fancy remote-controlled vehicles (such as the combat robots we've all seen on TV) to be robots unless the controller on the other end of the radio link is a computer. For example, I am building Zeppo as a mobile sensor platform for Groucho. However, a "remote-control" mode can be very useful in testing out parts of a robot. To be honest, other people have different definitions.

Types of robots

The behaviors and construction of a robot should be determined by the goals of the robot. For example, a robot for underwater observation needs different construction and programming than does a robot designed as a toy for a small child.

I tend to divide robots into different types, depending on their function:

- **Experimental robots (x-bots).** These robots are used to experiment with various facets of robotics.
- **Special-purpose robots.** These robots are designed for a single purpose, such as robot sumo or brain surgery.
- **General-purpose robots.** These robots are designed to work in a natural environment (such as a house, a workplace, or even outside) and perform a variety of tasks.

Linux Controllers

Since this book is about Linux robotics, I will be using Linux motherboards for my examples. Groucho makes use of a low-power (under 60 watts) PC-class motherboard (Figure 1-3). This motherboard uses the Mini-ITX form factor (around 7 inches by 7 inches). I use Gentoo Linux with the 2.6 kernel. I set this up almost as I would a server/desktop machine. Yes, I even have X on the system so that I can display some fancy graphics to give Groucho more emotional appeal (I haven't used this feature yet, but I will).

I also have a VIA M10000 motherboard. This is a 1-GHz machine that uses only about 20 to 25 watts of power. I used this board on another robot and may use it on a future robot.

Figure 1-3 Groucho's brain exposed: the iBase 890c Mini-ITX motherboard.

I chose to use the Gentoo Linux because of the ease of updating the system. Frankly, you can choose whichever Linux distribution you are most familiar with as long as Java 1.4 is available for that system (this isn't much of a problem; I use Blackdown Java, which is available as a Gentoo package). If you don't like Java, you can reprogram the examples in C or C++ fairly easily. The code and examples that I have in this book I am releasing under the Attribution ShareAlike 2.0 License 2.0 License, as stated at http://createcommons.org/licenses/by-sa/2.0/license.

Why Linux?

Most small robots are controlled by microcontrollers rather than by a full-fledged computer system. However, there are a few reasons one might want to use a Linux system:

- Linux is a multitasking operating system.
- Linux has high-level communications libraries.
- Linux has built-in networking.
- Linux has the capability for easy video and audio input and output.
- Linux has a full set of support calls and device drivers.

All of this implies a standard Linux system that is built like a small server. You also can make use of more resources, such as X, if you want to use graphics on the monitor.

I developed most of Groucho in isolation, mainly because there weren't a lot of projects out there when I started. However, now there are several other hardware/software projects available:

- **The Open Automation Project** uses Linux on a Mini-ITX board much like the one I have; its robot is even built on a Zagros Robotics base.
- **The LEAF Project** uses a laptop but runs Windows instead of Linux. It should be possible to run most of its software on Linux. One of my future projects is to port the LEAF software to Linux. One of the interesting things about this project is that the three main developers work in three different languages.

- **The $500 Linux Robotic Platform** is a project that is just starting but has some interesting ideas.

Robot Design 101

Now it's time to build a simple robot. I will build one also so that I can verify all the exercises in this book. The design principles are as follows:

- Type: x-bot
- Durability: medium
- Flexibility: high
- Ease of change: high
- Physical complexity: low

We want a simple experimental robot that is easy to modify and is able to take on different sensors. Since this is purely experimental, it doesn't have to be very durable.

We're going to build this robot out of either a simple kit or easily available parts. You can get most of these parts from the local building-supply store or hobby store and over the Web, if you'd rather build one than get a kit.

Only a few tools should be needed: soldering iron, drill, screwdriver, knife, pliers, saw, diagonal pliers, normal pliers, wire trippers, and crimper, along with a file or sandpaper (Figure 1-4). A multimeter will be necessary if you get into the electronics; I got mine for less than $20 at Radio Shack. I use a Dremal rotary tool for some of the drilling, shaping, and polishing. Screwdrivers and soldering irons are easy to find. I used a hand drill and drill press. The only tool I needed to buy specifically for robotics was the soldering iron (I would have had a soldering iron, but I had destroyed my last one by dropping a box on it). I even had the multimeter for normal electrical repairs

I have built two robots that used Linux motherboards:

- *Groucho* already has been described.
- *Stuart* is built around a 1/6-scale radio-controlled M5 Stuart tank from Wal-Mart. I used only the base, which contains the drive system. The tank was big enough to house a VIA Mini-ITX board and two 7-amp-hour SLA batteries. One battery is for the motors, and the other is for the electronics.

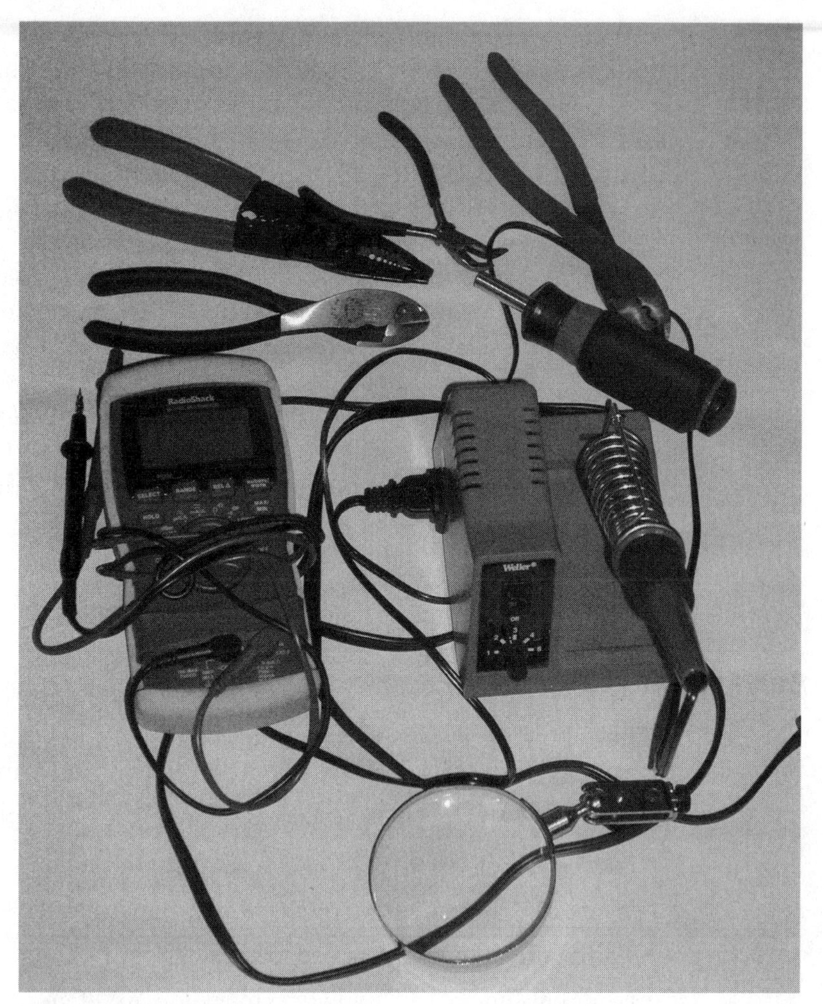

Figure 1-4 My basic tools.

You can get a fairly inexpensive prebuilt robot base from Zagros Robotics or can build your own. I think that it's easier and cheaper to go with the kit, unless you know what you're doing. The important things, from the point of view of this book, are that the robot should have a differential drive (that is, it has two wheels, powered by separate motors, and the steering, power, and braking are done by controlling the speeds of these motors) and that the motors can handle the weight (typically 30 pounds or more). I prefer a tank drive like the one on Stuart because I like the way it looks, but anything that applies to a differential drive applies to a tank drive (a tank drive slips a bit more than a wheeled differential drive and makes more noise, but that's the only major difference).

If you want to design and build your own base, I strongly suggest looking at the book *Build Your Own All-Terrain Robot*. Building a robotic base for even a Mini-ITX board can be challenging because the motors need to be heavy enough to carry the batteries that power the electronics.

You *do* want to make sure that your robot goes slowly enough. Yes, I said "slowly enough." It's extremely embarrassing to spend several hours or days building a robot only to have it self-destruct against the nearest wall. Even more embarrassing is to see it drive off the table and fall apart on the floor. And you really don't want to have your spouse yell at you for the damage to the wall and/or floor (or, worse, the dog; however, my dog is sensible enough to run away when he hears my robots coming). I'll admit that Groucho goes a bit more slowly than I'd like, but my house is too crowded for any real speed.

Building Robots

The following is a very basic discussion on how to build the physical parts of a robot. I suggest that you buy a robot base and modify it for your needs. Wheeled bases are available from

- Zagros Robotics: http://www.zagrosrobotics.com
- White Box Robotics: http://www.whiteboxrobotics.com
- Kadtronics: http://www.kadtronics.com
- Lynxmotion: http://www.lynxmotion.com. The 4WD3 base could be used if you employ energy-dense batteries, such as lithium-polymer batteries.

I'm sure that there are other sources, but these are the ones I've considered. I would recommend the Zagros Robotics base if you're comfortable with simple mechanical tasks and the White Box Robotics base if you aren't.

Building Groucho

Of all my robots, I've spent the most time on Groucho. However, this wasn't because it was the most difficult robot to build; rather, it was the first one I built using more robust techniques.

Figure 1-5 A deck direct from the factory.

The base was a Zagros Robotics 16-inch-diameter base with three extra decks. Each of these is made out of 1/4-inch black PVC. The decks are identical, with four mounting holes for placing supports for each deck. However, they are not exactly symmetrical, so you have to line them up carefully (Figure 1-5). I used 1/2-inch PVC pipe and four 3/8-inch (24 to 36 inches long) threaded rods to separate the decks. The bottom deck came predrilled with mounting holes for the motors and the casters.

In each deck, I drilled a large hole (2 to 3 inches in diameter) to pass wires through in the center of the deck. Then I cut four identical lengths of PVC pipe per level (Figure 1-6). I also put a large shoulder washer between the PVC and the deck to help support the weight. When I finished, I used a hacksaw to cut the threaded rods to length. Then I put an end nut (also called an acorn nut) on each end of the rod to keep things from falling apart.

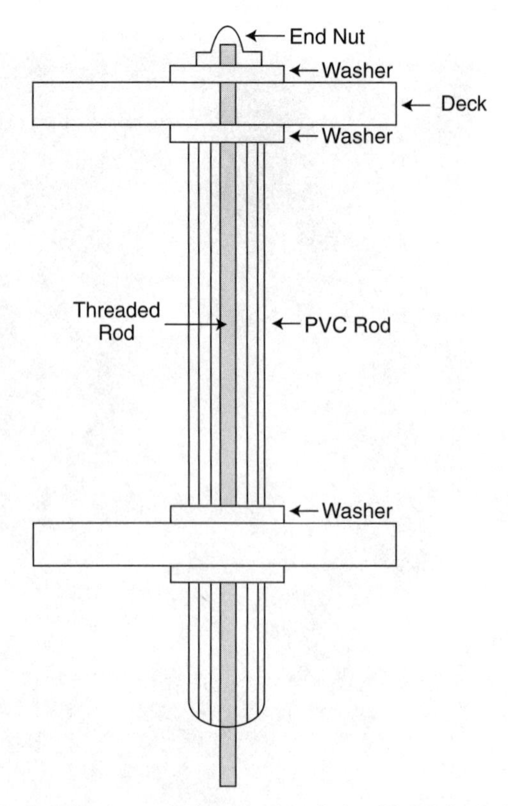

Figure 1-6 Half-inch PVC pipe and 3/8-inch threaded rod hold the decks together.

The motors (Figure 1-7) that came with this base are fairly powerful (300 oz-in) with built-in quadrature encoders for odometry.

To power Groucho, I have a 12-V, 12-amp-hour SLA battery for the motors and two 12-V 7-amp-hour batteries in series (for 24 V) run through a DC-DC converter to produce 12 V (Figure 1-8). The reason I do this is to extend battery life.

Groucho's main brain is an iBase 890c Mini-ITX motherboard with a 2.0-GHz Pentium M and 1-GB SRAM. I have a 40-GB laptop hard drive for storage (Figure 1-9).

The sensors are powered by a BDMicro MAVRIC-IIB with some custom programming. I use a USB to RS485 device to communicate with this board. The motors are controlled by a BDMicro REACTOR RX50 H-bridge prototype; these boards also are controlled by the same RS485 port.

The sensors include:

Figure 1-7 The wheel is to the right, and the encoder is built into the back (left side) of the motor.

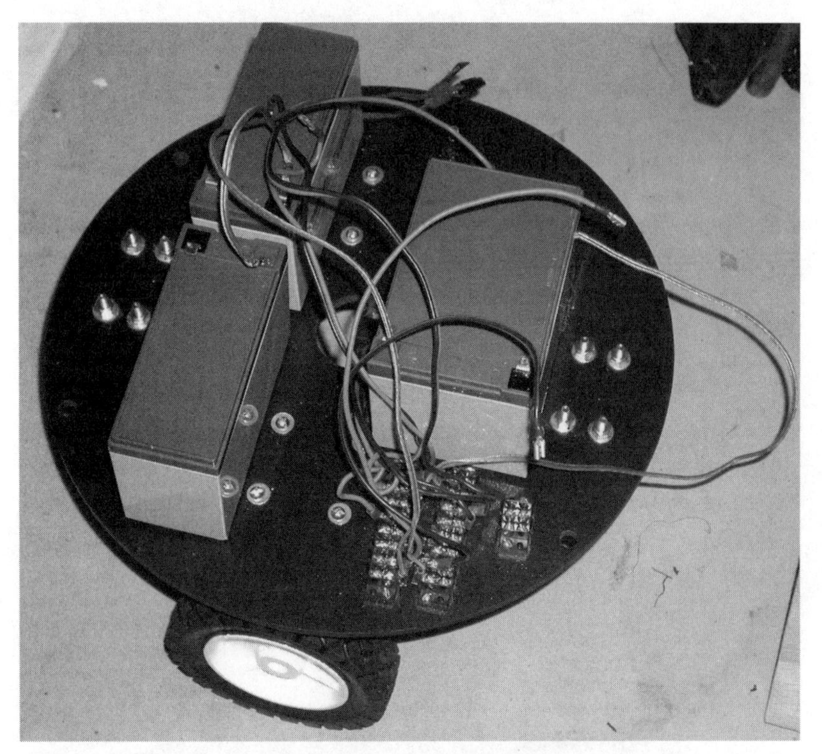

Figure 1-8 The first deck holds the batteries. The large battery powers the motors, and the small batteries power the electronics.

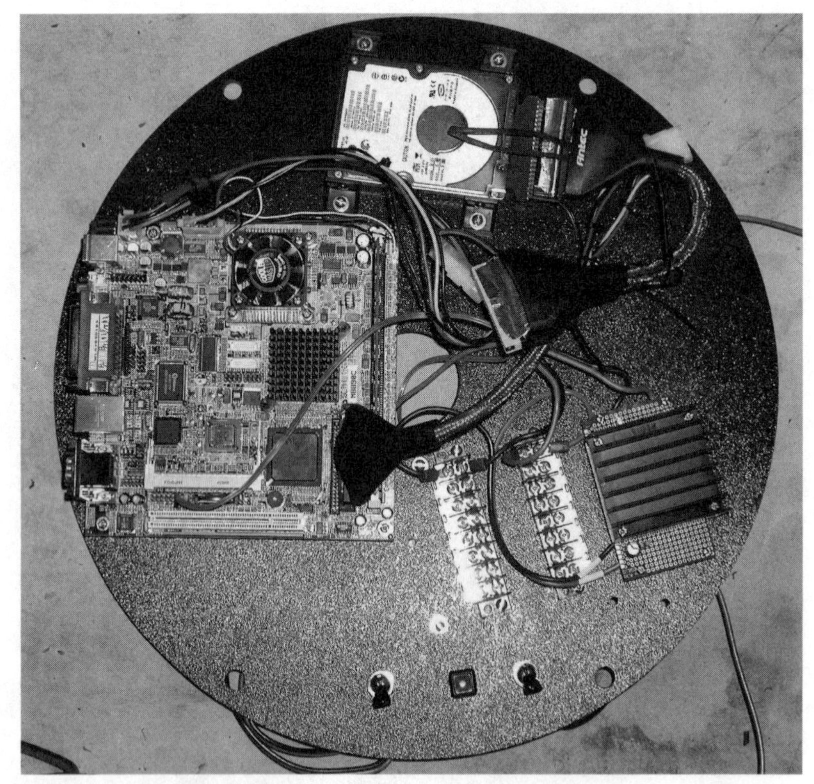

Figure 1-9 The second deck holds the motherboard and the power switches.

- Twelve Devantech SRF08 sonar units around the perimeter
- One Sharp IR Ranger for short-range sensing of dog toys
- Two voltage sensors for battery condition
- Video input provided by a Logitech Orbit webcam
- Audio input provided by a microphone
- Encoders attached to the motors for measuring speed and distance

The basic building techniques I used were as follows:

- Use barrier strips for most connections.
- Use some sort of connector for any wire that goes between decks.
- All wiring that goes between decks should go through holes in the decks.
- Use color-coded wire and be consistent about it.

- Use sturdy connectors (nuts and bolts) whenever possible.
- For objects that may be changed often, use 3M Dual-Lock.
- Fasten the sensors securely.

These may be sensible precautions for some people, but I'm not always sensible.

I have to say that I am impressed with the Zagros Robotics base. It is highly simple, but it works well.

Building a Robot Base

Unless you've built a robot before, I strongly suggest that you buy a base. Generally, the cost of buying the motors would be higher than the cost of an inexpensive kit. If you are good with tools, building a basic robot isn't very difficult.

Modifying a toy

If, for one reason or another, you don't want to build your own base and don't want to buy a robot base kit, you can modify a toy. Inexpensive radio-controlled (R/C) vehicles are the best for this.

Wal-Mart sells a 1/6 scale R/C M5 Stuart tank under its Motorworks label. I am using one of these for a different robot (Figure 1-10).

There are many other R/C tanks out there. Because of the size of the Mini-ITX format, I suggest you look at 1/8-scale or better.

Wal-Mart also sells some inexpensive R/C motorized treaded vehicles under the Motorworks label. There are both a tank and a missile launcher, both of which can be useful as a base if you use lightweight batteries.

With a little bit of work you can easily remove most of the electronics and substitute your own. In the case of Stuart (my M5 Stuart tank; see Figure 1-11), I removed both the receiver and the motor controller (which was based on relays). I changed the power supply from a 7-amp-hour sealed lead-acid (SLA) battery to two 9-amp-hour SLA batteries. I removed the inner walls to make everything fit, put the motherboard and batteries on the bottom, and used a plastic sheet to make a second level on which to put the sensors and other electronic. This made a sturdy base.

Figure 1-10 This toy makes a wonderful robot base.

Figure 1-11 Stuart: loud but good.

The Power Supply

For Groucho, this was easy. I used three SLA batteries: two for the Mini-ITX board and one for the two motors. Both the Mini-ITX board and the sensors have their own power regulators.

Even sensors can produce spikes; besides, most sensors use 5 volts, rather than the 12 volts that the motherboard needs. The exact power supply depends on the type of robot. Groucho was designed from the ground up to use small (7- or 12-amp-hour) SLA batteries (I did have to redesign the power system totally during the building process). If I wanted to build a lighter robot, I would use lithium polymer (LiPol) batteries because they are *much* lighter than anything else at the moment, but they are also much more expensive (LiPols have other problems at the moment). There are many other power sources that will be coming up in the future.

The power for the motors (Figure 1-12) comes directly from the batteries, only going through the H-bridge (the motor control circuit). The power for the electronics and the computer (Figure 1-13) goes through a regulated power supply; this takes the power from the batteries and makes sure that the power is the correct voltage. When I can, I use switching power supplies because they are more efficient and produce less heat than does a traditional linear power regulator.

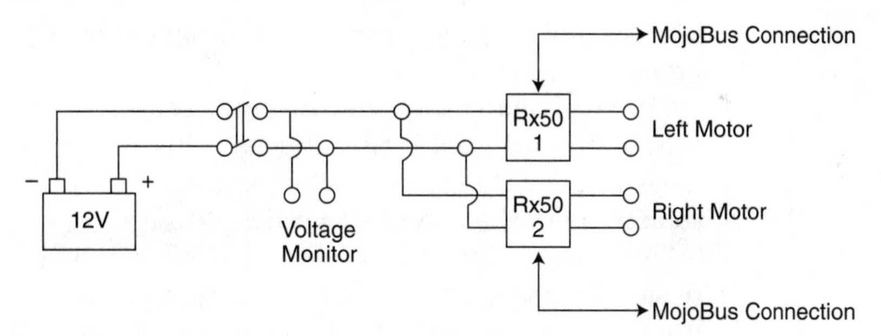

Figure 1-12 A block diagram of the motor system.

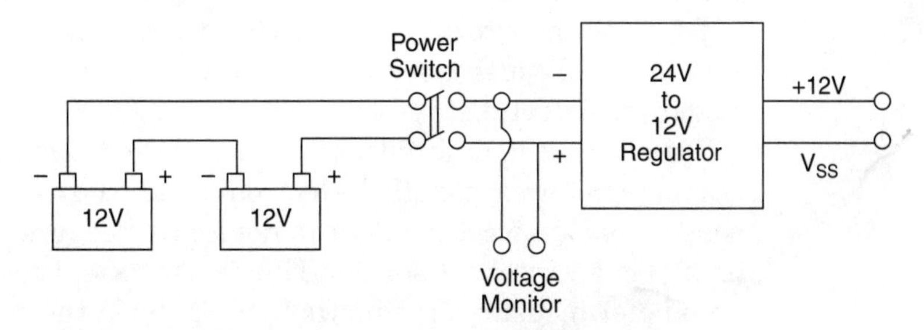

Figure 1-13 A block diagram of the electronics power system.

Groucho must have the batteries put into the bottom level, or the batteries will affect the center of gravity too much. He is designed so that the batteries fit on the bottom deck, near the main drive wheels.

It is possible to power a robot from pretty much any source of energy. You can use gasoline, solar power, or even clockwork. However, batteries are probably the easiest and cheapest system. I have a dream of making a steam-powered robot some day, though. Everybody needs a dream.

Sensors

If motors are the muscles of the robot, sensors are the eyes and ears and touch organs. You will have at least a couple of different types of sensors on your robots. Most of them will be mounted near the edge of the robot; this is why you want to leave that area empty.

A sensor is a device that takes readings of the world and converts them to a form that is readable by a computer. Most sensors that we'll be using are simple, in that they provide a single value for a given condition (Figure 1-14). Video and audio sensors provide much more information and will be considered later.

There are many ways to categorize sensors. For now, we will talk about internal and external sensors. Internal sensors measure something in the robot, and external sensors measure something in the external world.

Two types of internal sensors that we will be using are batter-level sensors to detect a low charge and odometers to measure the rotation of the wheels and thus to measure distance. Yes, odometers are internal sensors because they don't have a direct relationship with the external world. They measure the turning of a motor or wheel, which may be slipping or stuck.

Two types of external sensors that we will use are touch sensors (basically just glorified switches) and infrared (IR) range finders to detect distances.

Sometimes the differences between these things blur a bit: A magnetometer (a glorified electronic compass) is an external sensor, and an accelerometer (a device to measure acceleration or tilt) is an internal sensor. This is the case, because a magnetometer measures magnetic fields outside the robot, whereas an accelerometer measures the robot's acceleration or tilt.

Figure 1-14 An assorted group of common sensors.

One thing to remember: A robot is blind and deaf compared with pretty much any visible biological animal. The most fully loaded robot I've seen had no more than 50 sensors. Imagine walking around with most of your body numbed, wearing gel-smeared glasses and earplugs. Even then you would have more sensors than any robot does.

One corollary to this is that the more sensors a robot has, the more processing power is needed to gain information from those sensors. Vision is especially processor- and memory-intensive; audio is only a bit less so. The time the computer spends interpreting the raw sensor data is time unavailable for actually using the data.

Controlling Your Robot: The Software

The high-level control will reside in a Java application that runs on the main computer. However, since neither Windows nor Linux is a real-time operating system, we will use the serial ports to control things. I will connect the sensors either to the MAVRIC-IIB or to a Phidget Interface Kit 8/8/8 (a USB device).

From the point of view of the Java application, it doesn't matter how the robot actually is controlled. As long as the motors can be controlled and the sensors can be sensed, the exact means of doing this is immaterial. The hardware interface will be hidden by implementation classes.

There do need to be a few high-level interfaces for controlling the robot. For most purposes, the type of motor doesn't matter in regard to the programming; similarly, sensors are fairly similar from a programming standpoint. Unfortunately, from the controller's standpoint, things are different: A servo is controlled differently than is a gear motor, and different sensors may need to be handled differently. Therefore, I have a Motor interface but also have MojoMotor class to control motors controlled by the BDMicro RX50 (the working name for this controller was the MojoBridge).

Java handles this situation with interfaces. By using interfaces, you can, to some degree, ignore the hardware and concentrate on the software. However, I wouldn't ignore the hardware totally: Somebody I know is trying to control a (real) car robotically. A simple error with Groucho might result in a small mark on my wall; a mistake with a car can result in major damage.

The motor interface

```
public interface Motor {
    public static final Object FORWARD = new Object();
    public static final Object BACKWARD = new Object();

    // Motor initialization
    public void initMotor();
    public void closeMotor();

   // Motor power
   public void setPower(int newPower);

   public int getPower();

   public void setMaxPower(int newMaxPower);

   public int getMaxPower();

   // Keep the motor going with the same power, but in the
   // opposite direction; this can be dangerous with some
   // motors and geartrains
```

```
    public void reverse();

    // Set/get the current direction
    public void setDirection(Object newDirection);
    public Object getDirection();

    // Whatever the speed, put the motor into dynamic braking
    // mode and set the power to 0; if a motor doesn't support
    // dynamic braking, then just set the power to 0.
    public void brake();

    // Optionally supported commands

    // You need odometry for this
    public int getActualSpeed();

    public void resetPosition();
    public int getPosition();

    // PID Paramters
    public void setPID(int periodMs, double p, double i, double d);
}
```

The sensor interface

```
package ws.enerd.robots;

import ws.enerd.robots.sensor.SensorImpl;
import ws.enerd.robots.sensor.SensorListener;
import ws.enerd.robots.sensor.SensorFilter;

public interface Sensor {
    // Return the main value of the sensor as an integer or object
    public int getValue();
    public Object getObjectValue();

    // The interval in milliseconds between reads
    public void setReadInterval(int newReadInterval);

    public int getReadInterval();

    // An active sensor is somehow broadcasting; some sensors need to be
```

```
    // activated before having their value read
    public void setActive(boolean active);

    public boolean isActive();

    // Handle the case of a filtered sensor
    public void addSensorFilter(SensorFilter newFilter);

    public void removeSensorFilter(SensorFilter newFilter);

    public void filterSensor();

    // Deal with the SensorListener's
    public void addSensorListener(SensorListener listener);

    public void removeSensorListener(SensorListener listener);

    // Set the value and if changed call fireSensorListeners
    public void setValue(int newValue);

    public void setObjectValue(Object o);

    public void fireSensorListeners();

    public boolean shouldRead(long time);

    public void readSensor(long time);

    public void setSensorImpl(SensorImpl impl);
}
```

Linux and the Mini-ITX Board

By the time this book sees print, there probably will be smaller and lower-power designs. VIA has a Nano-ITX form factor that is about 4 by 4 inches, lacking only a standard serial port. I would have used a Nano-ITX board, but they hadn't come out when I started this book. They still haven't come now that I'm finishing it.

There are other ways to use Linux in a robot:

- **PC-104 form factor boards.** These boards follow an industry standard for embedded computers. They are typically easier to interface with the outside world, but they are also more expensive and typically are slower.

- **Single-board computers (SBCs).** There are a number of very small Linux boards that are based on various chips. They typically require much less power than the boards I chose, but they have less memory and are slower. After I finish this book, I'd like to explore some of those boards. They are typically smaller and slower than the boards I'm using for this book, but sometimes you don't need the speed. An example of this type of computer is the Gumstix board, which uses very little power and runs Linux reliably. Maybe by the time I'm finished with this book the SBCs of this class will have the resources of the Mini-ITX boards. Even if they don't, I may try to use a small network of them, with each Gumstix handling a different task.
- **Full-scale motherboards.** These are for large robots that can carry large batteries. I'm not building anything that large for this book, but I have plans to build one in the future. The main advantage to using a full-size motherboard is that you have greater expansion capabilities.
- **Laptops.** Some robots are built to carry a laptop computer that is the main controller. I'm not fond of this arrangement because a laptop carries additional weight that a robot doesn't need. However, used laptops are fairly inexpensive. The LEAF robots are built around laptops.

Storage

Groucho uses a 40 GB laptop hard drive. This requires less power than a traditional hard drive and can take more shaking around.

Another alternative is to use Compact Flash cards as hard drives. They are nice because they use very little power (about half a watt) and are physically strong. Unfortunately, they are expensive. I do have a 4-GM hard drive that is built in a CF card; it is large enough to install Gentoo Linux.

I chose to put Gentoo Linux (http://www.gentoo.org/) on this board. This was a bit different for me, because until then I mainly had used RedHat Linux. I chose Gentoo mainly because it was extremely configurable and extremely easy to upgrade.

If you opt to use Compact Flash cards for your main storage, you probably won't be able to put a full development system onboard (well, I wasn't; perhaps CF cards will get bigger shortly). My solution was to mount the partitions that hold the source code via the Network File System (NFS). This means that I can have access to multiple gigabytes of source code while updating my system without having to have a huge amount of storage on the robot itself. You can do this without using NFS by having a removable hard drive; a USB hard drive will do.

Putting Gentoo Linux on the Mini-ITX

Gentoo has an excellent installation guide. Although the installation isn't as easy as it is for RedHat/Fedora, it isn't difficult. The handbook is fairly step by step and easy to follow. I suggest that you look at the Gentoo site for more specific and more current details.

Additional software

As optional software you will need a few things:

- Java: I use Blackdown Java 1.4.2.
- The Java serial classes, available from http://www.rxtx.org: I use these same classes on my Windows machine instead of the Sun classes, so that I can stay consistent with my laptop and robot.
- Festival for text-to-speech output. It is available from http://www.cstr.ed.ac.uk/projects/festival.
- Sphinx4 for speech-to-text input from http://cmusphinx.sourceforge.net/sphinx4.

Everything except Sphinx4 is available as a Gentoo package that works with the Gentoo portage system. You shouldn't have much trouble getting things to work.

Summary

- By the time you finish this chapter, you should have enough information to choose your robot base. You will need the base in the rest of the book.

- If you don't like simple bases, you may have to make some adjustments in the exercises later in the book. If you convert a toy, you'll have to deal with the particulars of your base.
- If you are new to building robots, I strongly suggest that you buy a base rather than trying to figure out the details of motor placement and power distribution.

Chapter

2

Electronics and Such

Do not fear: This isn't a textbook about electronics. Although there will be sections that require a basic knowledge of circuits and how to build them, I won't deal with anything difficult. Frankly, I built only two boards in my robot: One was a 12-V power supply for the electronics, and the other was a 5-V power supply and I2C breakout board.

The concepts in this chapter are as follows:

1. The ground of electronics: electricity
2. Motors
3. Basic components
4. Schematic diagrams
5. The soldering iron and friends

If you want anything more than the simpler overview that is in this chapter, you'll have to get some other books to supplement this material. There are several reasons I don't want to write a lot about electronics:

1. The more I write about the basics, the less I can write about robotics.
2. My publisher has several electronics books available, and you can get one of them.

Electricity

The most basic thing is that electricity exists and, as far as this chapter is concerned, flows from positive to negative. I'm not going to worry about some of the details that concern physicists. If you want to get into that level of detail, please look elsewhere. There are many good books on theory that I use as paperweights. If you know enough theory to argue with me, you probably don't need to bother with this chapter.

Ground (V_{ss})

The zero voltage of the system is called the ground. Yes, this used to be connected to the physical ground with a large metal stake and sometimes still is. However, that would be somewhat impractical in robotics, and it doesn't matter anyway for what we'll be doing.

For historical reasons that I will make no attempt to explain, ground usually is called V_{ss}.

Since ground is the zero voltage, it is important that all the parts of a circuit share the same ground. This isn't difficult to do if you plan it, but if you accidentally create two slightly different ground voltages, you will have created what is known as a ground loop or, in technical terms, a mistake. This can cause all sorts of problems, ranging from buggy communication to fried electronics.

Voltage In (V_{in})

I'm going to simplify things here. Most simple circuits, such as the ones we'll be discussing in this book, have a single input voltage, which normally is called V_{in}, V_{dd}, or V_{cc}. The main difference among them is that V_{in} is typically an unregulated battery voltage, whereas V_{dd} and V_{cc} are (one hopes) regulated and stable.

V_{in} is typically the highest voltage in the circuit, but not necessarily. However, for now think of it as the highest normal voltage and you'll get the idea as you go along.

For the purposes of my example robots, V_{in} is 12 volts, with only one exception.

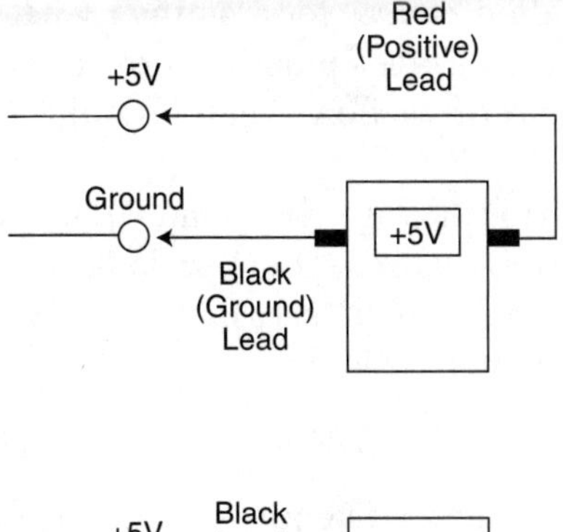

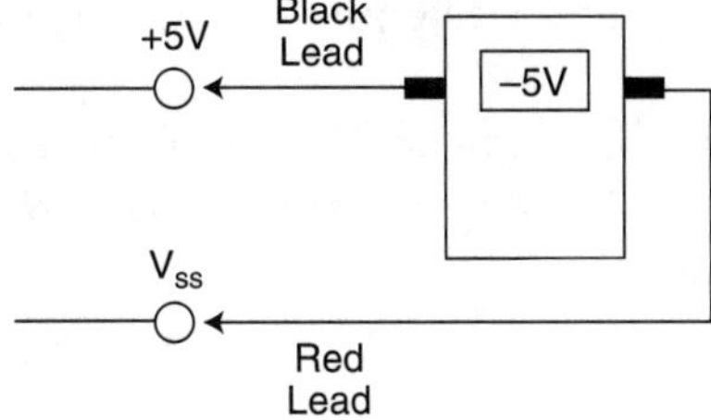

Figure 2-1 The voltage difference is either +5 volts or –5 volts, depending on how you measure it.

Positive/negative

I mentioned positive and negative earlier in the chapter. For most of the circuits in this book, the positive value will be V_{cc}, which will be +5 V. The ground value (V_{ss}) will be, by definition, zero volts (0 V). Therefore, V_{in} is positive with respect to ground, and ground (V_{ss}) is negative with respect to V_{in} (see Figure 2-1).

For example, if you measure from V_{cc} to V_{ss} with a voltmeter, you should get (for my purposes) 5 V. If you reverse the leads of the voltmeter, you will get –5 V.

Current

Current, or the flow of electricity, is measured in amperes (amps). If voltage is the potential between two points, current is the actual flow.

Resistance

Resistance is a property of any substance. Things with low resistance, such as copper, iron, and many other metals and some plastics, are called *conductors*. Things with high resistance, such as rubber and most plastics, are called *insulators*. Conductors allow current to flow freely. Insulators block the flow of current.

The more interesting things lie somewhere in between. Resistance is measured in ohms (Ω). Ohm is best known for his definition of resistance, which is called as Ohm's law. It is explained by the simple equation:

$$e = i \times r$$

This means that voltage (e) is equal to current flow (i) times resistance (r).

If you're wondering about the odd letters for voltage and current, it helps to think of them as arbitrary or to realize that Georg Simon Ohm was German.

Voltage drop

Voltage drop is one of the main concepts in electronics. Two points that are directly connected have the same voltage.

If those points have different voltages, there is a short circuit. This is usually a bad thing and should be avoided. The last time I did that, I vaporized a connector. At the voltages and currents we will be using, the worst you'll get should be a minor shock and the destruction of parts of your circuit. Before you start using circuits with high voltages and currents, please learn a lot more about electronics—much more than I know, for example.

If the two points have a component with resistance between them, each point will still have its original voltage and the component will have current flow determined by Ohm's law.

The voltage across this component is called the voltage drop (Figure 2-2).

Power spikes and transients

In an ideal world, a 12-V supply would remain at 12-V as long as there was enough power. However, in the real world, power doesn't flow that smoothly.

Power supplies fluctuate as the power is used. If there is a sudden demand for power, the supply will drop momentarily. If power is fed back into the circuit, the power will rise momen-

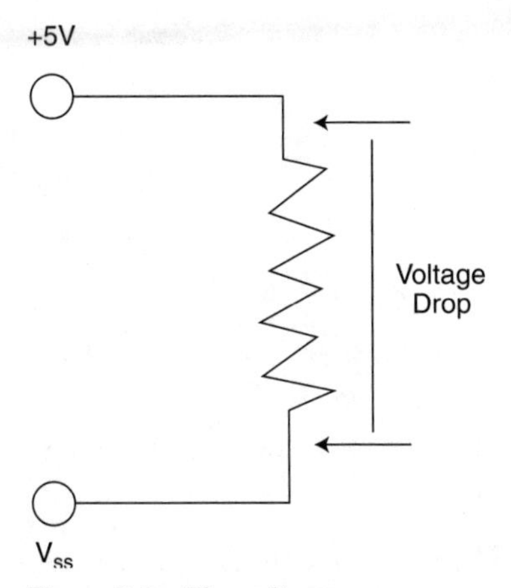

Figure 2-2 The voltage across a component is called the voltage drop.

tarily. These events are called *transients* or *power spikes*. Motors are a common cause of power spikes because they need a lot of power for a short time while they are turning on and actually can generate current when you slow them down quickly. This is why I always have separate batteries: one for the motors and the other for the electronics.

However, even electronics have power spikes. One of my favorite sensors, the Sharp IR Ranger, uses up to 400 milliamps (mA) momentarily around 400 times per second. This can cause enough of a power spike that it will reset microcontrollers and generally have bad results. This is why I take care around these components.

Motors

Every robot book has to talk about electric motors; it's an unwritten rule. Therefore, I'm going to go over them very briefly; if you want to learn more, you can pick up any of a number of books on motors (I suggest reading *Building Robot Drive Trains* by Dennis Clark and Michael Owings for more information). For my part, I don't care about the inner workings of motors except in regard to how they affect my robots. To me, a motor is a black box that takes electricity in and produces rotation and noise (here, noise means both unwanted sound and unwanted electrical interference). However, there are a few basics that any robot-maker should know.

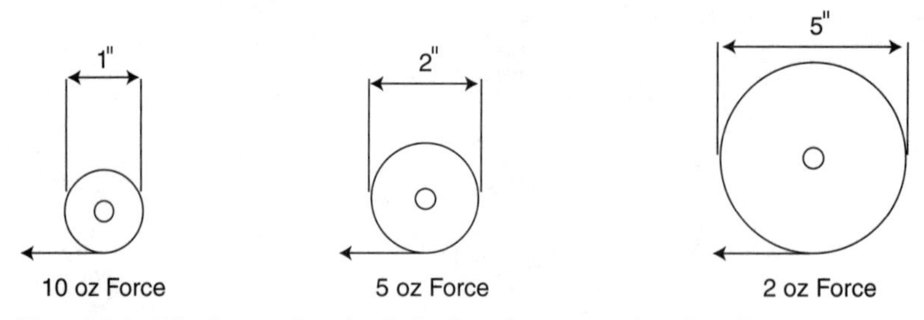

Figure 2-3 The larger the wheel, the less force is produced at the rim.

- **Nominal voltage.** This is the voltage for which the motor is built. You sometimes can use a higher voltage and get more power out of the motor. This can destroy the motor.
- **RPM** (revolutions per minute). There are a couple of variations: RPM under no load and RPM under load. RPM under load is actually useful.
- **Torque.** This is twisting force produced by the motor (Figure 2-3). It is measured in units of length and force, such as inch-ounces or foot-pounds. A quick description is that a motor with 1 foot-pound of torque can, when used with a 1-foot-radius wheel, produce 1 pound of force at the edge of the wheel.
- **Mounting.** You need to know how to mount the motor to the frame and the wheels. Different motors have different mounting holes. This is where a drill comes in handy.

Permanent magnet DC motors

This is the basic electric motor and generally is called a DC motor or a gear motor. It has two leads, and when you supply power, the motor shaft spins (Figure 2-4). If you supply less voltage, the motor spins more slowly; if you supply more voltage, the motor spins faster; if you supply too much voltage, it overheats, may spark, and eventually goes to la-la land.

Most DC motors spin too fast to be useful in a robot as is. Therefore, many of the motors you'll see come with a built-in gearbox (hence the name *gear motor*). The basic rule for gears is that if the RPMs decrease, the torque goes up proportionally, and vice-versa. There will be frictional losses, but that's the basic rule.

Pulleys work the same way as gears in this respect.

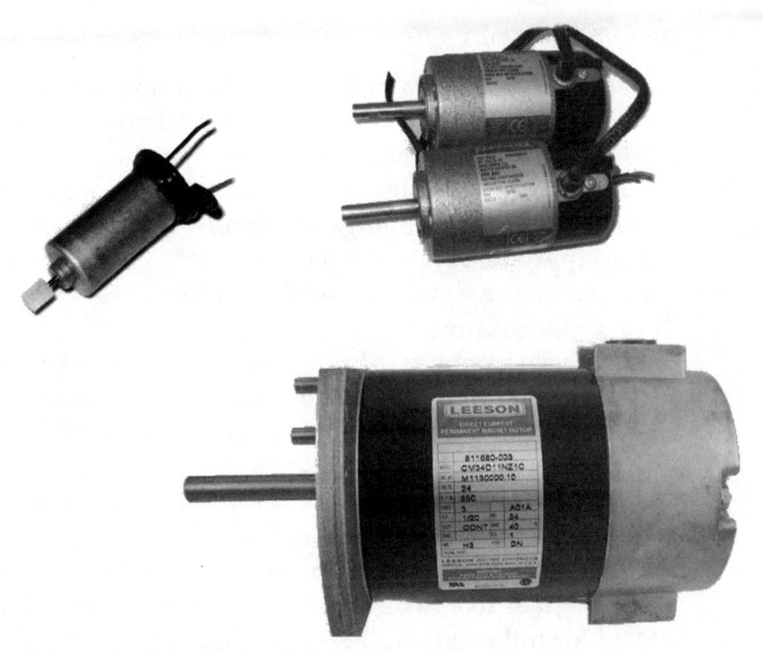

Figure 2-4 Some random DC motors.

Pulse width modulation

I said before that the speed of a motor can be controlled by the voltage. However, it is fairly inefficient to do this directly. Luckily, there is a good digital solution that fits well with microcontrollers: pulse width modulation (PWM).

PWM works by keeping full voltage on only a part of the time; the rest of the time the voltage is 0. If full power is desired, the power is on the entire time. This is done quickly enough so that the motor keeps running. If we assume a frequency of 1 KHz (1,000 cycles per second) and you want to have the power controlled so that the power levels are integers from 0 to 8, for full power (level = 8) the voltage is on for the entire millisecond. If the power level is 4, the voltage is on for the first half of the millisecond and off for the second half. If the power level is 1, the voltage is on for the first eighth of the millisecond and off for the rest. The percentage of time the power is on is called the duty cycle.

Typically, you will use a circuit called an H-bridge to take the PWM from the controller and the motor input voltage to power the motor. An H-bridge is so named because the schematic diagram looks somewhat like the letter H.

R/C servo motors (servos)

Radio control servo motors (servos) typically are used in the R/C model community but have branched out into robotics. A servo is a gear motor with the speed-controlling circuitry already attached and usually some hardware to prevent the servo from rotating fully (originally, a servo was designed to move parts of a model back and forth). With a simple modification (you can buy premodified servos) a servo can rotate like a normal gear motor.

A servo requires three connections: power, signal, and ground. The signal should be at +5 V when on and 0 when off and can come directly from a microcontroller. The power is generally between 4 and 6 volts.

Controlling the speed of a server is a bit different from controlling that of a gear motor. The speed controller of a servo is designed just to control the position of the servo from about –90 degrees to +90 degrees (Figure 2-5). When a servo is modified for full rotation, the speed controller controls speed.

A servo expects a different sort of PWM than does a gear motor: A pulse of about 1.5 ms is sent 50 times per second to the servo. For a normal servo, a pulse of 1.5 ms puts the servo in its center position. A pulse between 1 and 1.5 ms moves the servo somewhere to the left in proportion to the

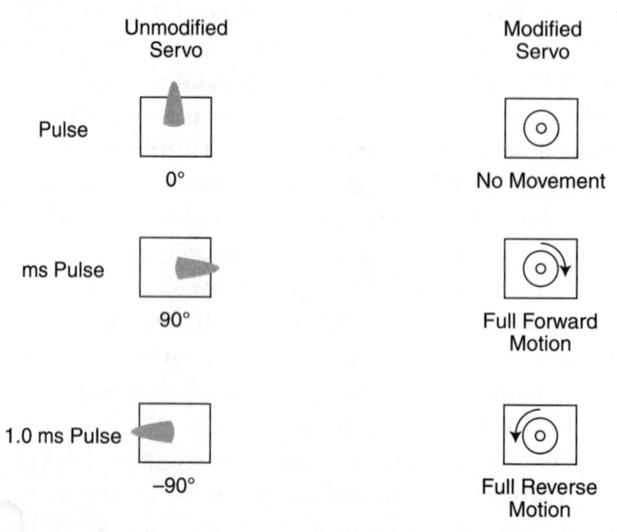

Figure 2-5 An unmodified servo goes to a specific angle; a modified servo rotates in that direction faster as the angle will increase.

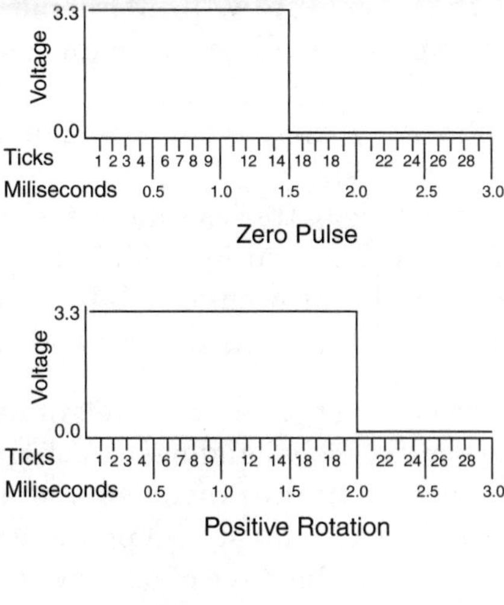

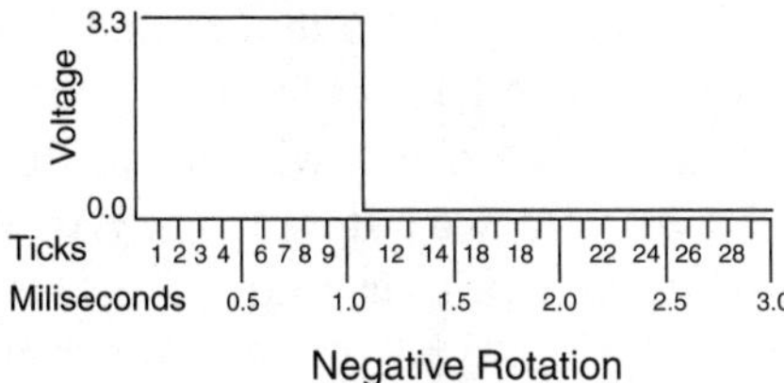

Figure 2-6 Servo pulses.

time; a pulse between 1.5 and 2 ms moves the servo in the opposite direction.

To control a modified servo, all you need to do is to use a pulse between 1.5 and 2.0 ms for forward (the longer the pulse, the faster the movement) and a pulse between 1.0 and 1.5 ms for backward (the shorter the pulse, the faster the movement), plus a pulse of 1.5 ms for no movement. Most servos tend to remain in the last position if you stop the train of pulses, and so you can use this for braking also (Figure 2-6).

There is a device called an electronic speed controller (ESC) that takes a servo signal and converts it to a PWM signal that can control a normal DC motor. I have used ESCs to control one of my robot's drive motors.

Batteries

Our robots will need electrical power, and this is the whole reason for talking about electricity and electronics in the first place. For robots to be mobile, they must carry their power with them. Until they come up with affordable fuel cells or nuclear generators, batteries will be the main source of power for robots.

Batteries are defined by their type, their nominal voltage, and their rating in amp hours.

The types of rechargeable batteries that are commonly available are sealed lead acid (SLA), nickel-cadmium (NiCad), nickel metal hydride (NiMH), lithium-ion (LiIon), and lithium-polymer (LiPol) batteries. Each has its pros and cons.

- **SLA** batteries have low energy for their weight and size; however, they are very inexpensive and easy to charge. They are one of the first types of batteries invented and are very simple. Basically, they have a group of lead plates separated by a liquid electrolyte solution. One type of battery that I will refer to as an SLA is gel-cells, which have a gelled electrolyte and are better able to withstand leaks. One problem with most SLA batteries is that they do not withstand deep discharge (i.e., being drained all the way down) very well. I strongly suggest the use of wheelchair or marine batteries, because they are designed for this type of deep discharge. Another problem with SLAs is that, as the battery discharges, the voltage goes down. A 12-V SLA is considered discharged at 9 V, which is too little power for the electronics of my robots.
- **NiCad and NiMH** batteries are about midrange, but they still cost a lot compared with SLAs.
- **LiIon and LiPol** batteries have a very high amount of charge for their size and weight, but they cost a small fortune. You can draw a lot of current from them quickly. However, if you draw too much, the battery can overheat, causing a thermal event such as setting your robot on fire or even creating an explosion. Although I am fond of explosions, I don't like them happening in the middle of my robot or my house. Also, the chargers for these batteries are more expensive than the chargers for other batteries.

For a Mini-ITX board you generally will need a battery with a voltage of 12 V. However, the voltage specified on the battery is the *nominal voltage*. For example, an SLA with a nominal voltage of 12 V is fully charged at 13.5 V. A NiMH cell nominally at 1.2 V has 1.4 V when fully charged. You must take this into account when you design circuits.

Whereas the voltage tells you what sort of circuit the battery can power, the *amp-hour rating* tells you how long the battery can power it. An amp-hour is just that: how many hours a battery can provide 1 amp. Well, sort of; the rating normally is calculated as how many hours the battery can power a 1/10-amp load divided by 10. I find that it's best to use the amp-hour rating as a way to choose between batteries of the same type and brand.

Another important characteristic of a battery is its discharge curve. This allows you to predict the behavior of a battery under load. Whereas NiCad, NiMH, and lithium batteries have a fairly flat discharge curve, SLAs start at 13 or more volts and are considered discharged (or empty) at around 9 volts.

For small robots you can use small numbers of nonrechargeable batteries, but to power a Mini-ITX board I find that small SLA wheelchair batteries are the best bargain. However, the pricing may change by the time you read this book, and the fancier batteries may have come down in price enough to be usable. Or, you could be willing to spend more money on batteries than I am.

Basic Components

Most electronic devices are composed of simple parts. The simplest description of an electronic component is that it is a device for converting electricity into heat (this is an unintended effect in most cases).

It is also well known that the active part of any electronic component is smoke, because once you let this smoke out (by connecting the part to a high-voltage source), the part no longer works.

Resistors

A resistor resists current flow. That's the long and short of it. Every substance has resistance. Conductors (such as copper) have very low resistance. Insulators should have high resistance. Resistance is measured in ohms and is defined by Ohm's law.

Resistors (Figure 2-7) used to be made of a length of high-resistance wire wrapped around a core. Today, they are made of many materials, including carbon film.

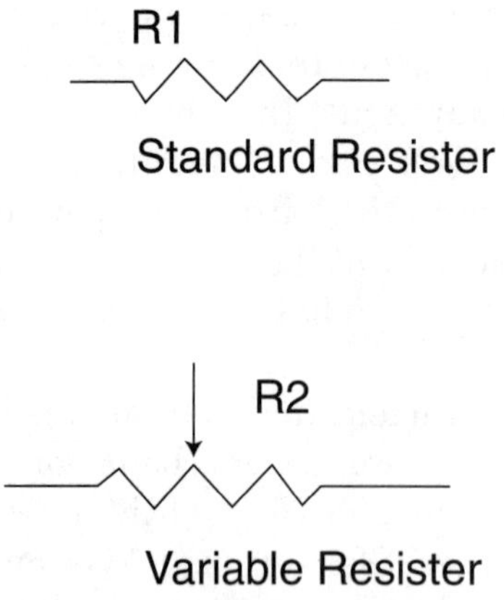

R1

Standard Resister

R2

Variable Resister

Figure 2-7 Resistor symbols.

Resistors can be fixed or variable. Most variable resistors are turned by a knob or screw and are called pots (potentiameters).

The uses of a resistor for the purposes of this book include the following:

- **Current limiting.** A resistor can be used to limit the current that will pass through another device. For example, if you want limit the current to 20 mA at 5 V, you can use Ohm's law to determine this resistor: $r = e / I$ = 5/0.02 = 250 Ω. Any component connected in series with this resistor will never have more than 20 mA running through it.
- **Voltage dividers.** If you have two resistors connected in series with a tap between them, you can put the voltage at the tap anywhere between V_{ss} and V_{cc}. This is done by using Ohm's law and basic algebra. If V_{cc} is 5 V and you want 1 V at the tap, you put the top resistor at 400 Ω and the bottom resistor at 100 Ω.

Capacitors

A capacitor stores voltage for a while. In some very small circuits you can use capacitors in place of batteries for short-term use. The main use of capacitors in the circuits in this book is to smooth out voltage fluctuations (Figure 2-8).

Between V_{cc} and V_{ss} a relatively large capacitor and a small capacitor will be placed. This shields the circuit against large

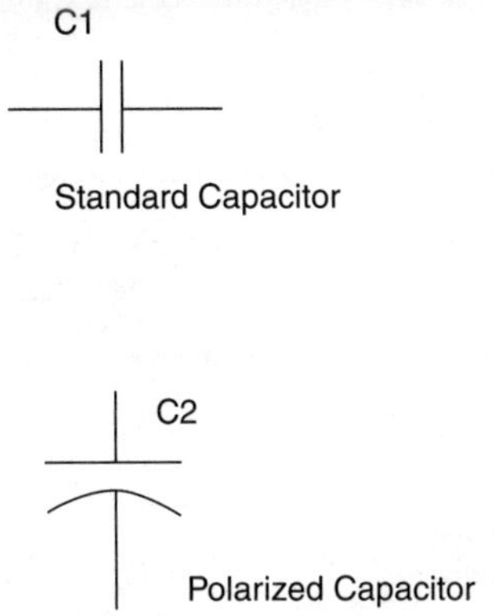

C1

Standard Capacitor

C2

Polarized Capacitor

Figure 2-8 Capacitor symbols.

voltage spikes in the power supply. Two capacitors are used to protect against both high-speed and low-speed fluctuations.

Between the V_{cc} and the V_{ss} of each chip there should be a very small capacitor. This capacitor shields the chip against high-frequency voltage transients. It is called a bypass capacitor. Strangely enough, these capacitors are not always shown on circuit diagrams.

Another common use of capacitors is to create a simple oscillator (a circuit that produces a voltage that goes up and down in a predictable way). This also requires a resistor and is called an RC oscillator.

Voltage regulators

A basic linear voltage regulator is a simple device. V_{in} is on one pin, V_{ss} is on the middle pin, and V_{cc} is on the output pin (Figure 2-9). If you put a few capacitors on the output, you will have a constant voltage power supply.

The one problem with linear regulators is that any excess voltage is turned into heat and wasted. However, these regulators are inexpensive and easy to use.

They can be fairly efficient also if the V_{in} is close to V_{cc}. For example, if V_{in} is 6 V and V_{cc} is 5 V, the efficiency is 80 percent.

A more complex voltage regulator is the switching regulator. These regulators generally require a few more parts, but they efficiently change V_{in} to V_{cc} with little wasted heat. Some of them run at greater than 95 percent efficiency. However, they

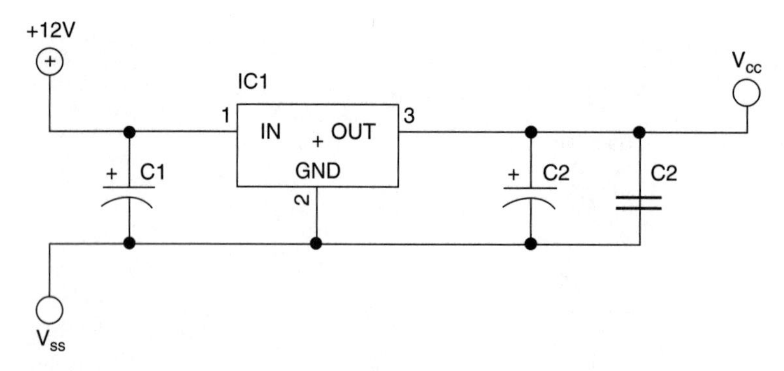

Figure 2-9 A simple voltage regulator.

have a big problem in that they cost more than linear regulators and require more parts. However, less heat is generated from a switching regulator.

Schematic Diagrams

An abstract diagram of a circuit is called a schematic diagram. It is composed of the symbols for the various parts, their values, and the connections between the parts. A good analogy is that a schematic diagram is very much like a road map for electricity.

To somebody skilled in electronics, a schematic diagram is enough to build a large and complex circuit. In this book, I will keep the schematic diagrams simple.

Breadboards, Printed Circuit Boards, and Soldering Irons

Now that you've gotten this far, you get rewarded by doing some actual work. Although the schematic diagram is enough to show you what the circuit is, you sometimes need to build a real circuit. There are many ways to build a circuit.

Soldering

I didn't want to include this part because this is not an electronics book; there are lots of those on the market. However, some of you might be true beginners and want a quick overview. Everybody else is free to disregard this section.

Solder is a metal that is designed to hold stronger pieces of metal together. It was used in plumbing and metalwork long before electronics came along. In the old days, solder was made from a lead-tin alloy. In the United States, at least, those days soon will be over, and solder will have to be made lead-free.

A soldering iron (sometimes called a soldering pencil) is basically a straight stick that has an end that gets hot enough to melt solder. It's actually a bit more complex than that: The base usually is molded so that it can fit well in one hand, and there is generally a cord coming out of the bottom. You don't need an expensive soldering iron, but you should get a decent one.

The basics of soldering are the following:

1. Turn on the iron.
2. Once it is hot enough, tin the soldering iron tip (melt a bit of solder on it).
3. Clean off the excess solder with a damp sponge.
4. Heat the joint with the soldering iron for 5 to 10 seconds.
5. Apply solder to the joint but not directly to the soldering iron.
6. Keep the iron on the joint for a few seconds until the solder has saturated the joint completely.

The job of a soldering iron is to get hot enough to melt solder, which melts at low temperatures. When I say low, I still mean hot enough to give you a very bad burn.

Some safety notes:

1. When I say that solder is a metal that melts at a low temperature, please pay attention to the words *metal* and *melts* rather than the word *low*. If you get molten solder on your skin (and if you're like me, you will), it hurts.
2. Do not, under any circumstances, hold onto the soldering iron by the hot end.
3. Do not let the hot end of the soldering iron get into your work area by accident. Very bad things could happen.
4. Dropping a soldering iron on a table or rug is a bad idea, not to mention a safety hazard.

5. Do not accidentally hold your arm or hand above a hot soldering iron.

Basically, the key to soldering is to put the soldering iron on the joint close to where you want to solder. The soldering iron heats up the joint where you will apply solder. When the joint is hot enough—in a few seconds—you apply the solder to the connection.

Printed circuit boards

A PCB (printed circuit board) is a thin board on which circuit traces are printed in copper (Figure 2-10). This allows one to solder electronic devices directly to the board without using wires. Most computer motherboards are built on PCBs.

You can make a PCB at home by using a variety of techniques, or you can have one built to order for a relatively low price.

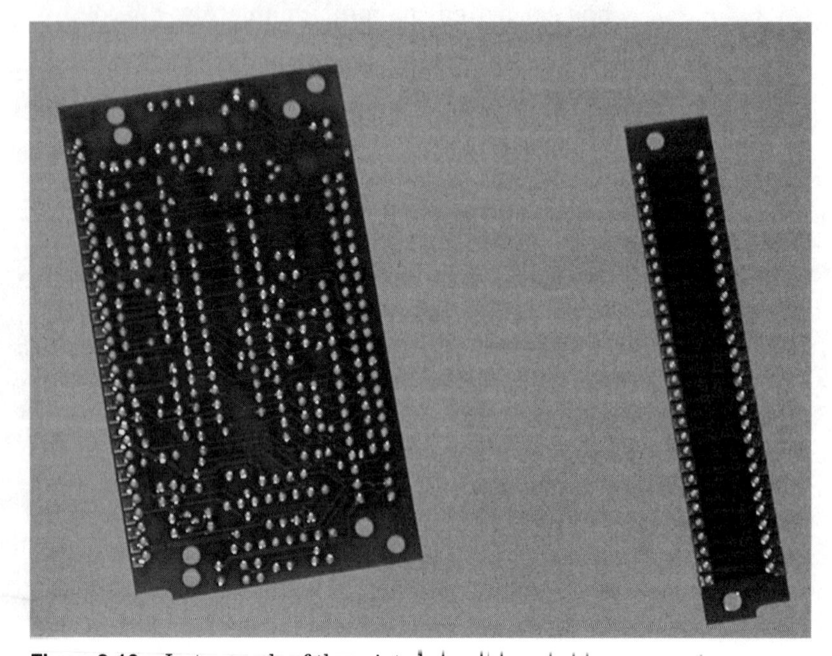

Figure 2-10 Just a couple of the printed circuit boards I have around.

Breadboards

A breadboard is simply a generic PCB to which you can solder pretty much anything (Figure 2-11). Both of the boards soldered for Groucho were soldered on breadboards.

Breadboards were the mainstay of experimenters for years. Unfortunately, as new parts get smaller and smaller, breadboards are not as useful as they once were.

Connectors

Perhaps the most time-consuming thing I did when I made Groucho's hardware was make the connectors. Often, there are no premade connectors for what you want to do.

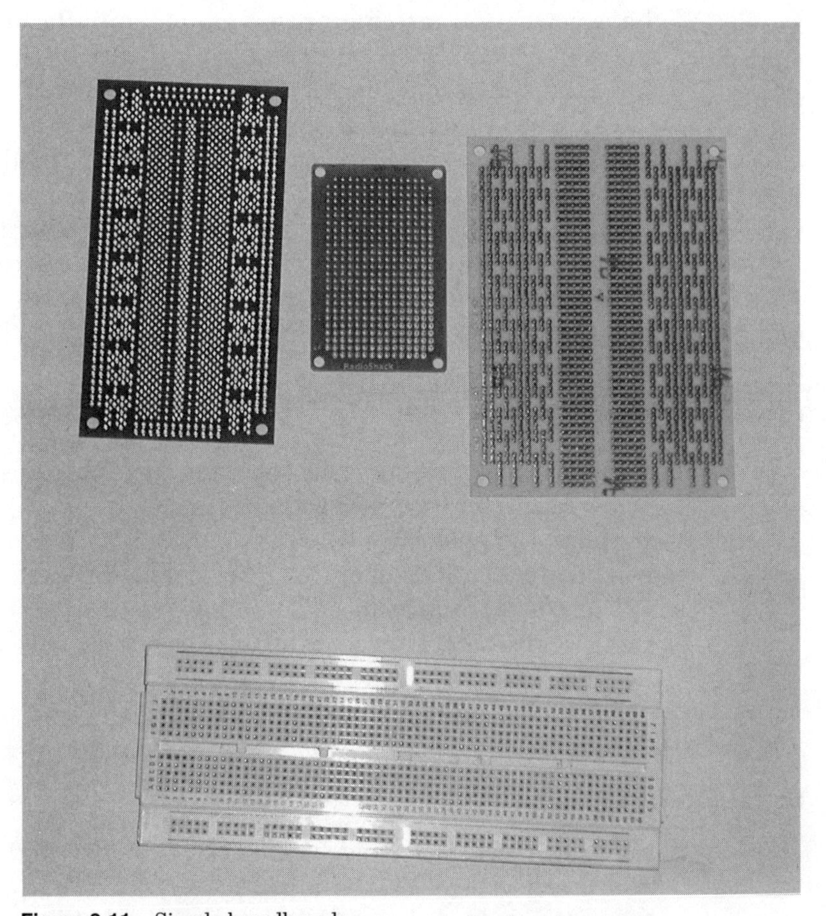

Figure 2-11 Simple breadboards.

I used three types of connectors for this project: 0.1-inch pin headers, barrier strips, and wire screw connectors.

Pin headers

Pin headers are just that: pins spaced (in my case) 0.1 inch apart in a straight line. These pins are easy to put on a board. They come in various lengths, and you can break off what you need and solder them to your PCB. These are called *male headers*.

Female headers are more difficult to find and make. In fact, I found the best connectors only a couple of months ago. I am referring to AMP MTA100 connectors (Figure 2-12). I use the polarized female IDC (insulation displacement connector) pass-through connectors with five pins. I bought the AMP T-Tool for putting in wires. If I had thought it out, I also would have gotten the matching polarized male pin headers so that my connectors would have fit together more securely. I also have the strain-relief caps for these connectors.

The reason I used pass-through connectors was that I could quickly make a single cable that would attach to all my sonar units. That meant less work. The most important thing when using these connectors is to make sure you have the correct size wire for the connector. This is listed with the connector in the data sheet for the connectors.

I bought these connectors from DigiKey.

Barrier strips

Barrier strips are very simple. They are plastic strips with screw connectors on them so that you can connect opposite connections easily (Figure 2-13). There are also metal strips you can buy for connecting one or more channels.

There is a crimp connector to allow wires to be screwed into place. I also solder these crimped wires to be sure of the connection.

I get these barrier strips at Radio Shack.

Screw connectors

Screw connectors are blocks that have an opening on the side and a screw head on the top (Figure 2-14). You put stripped wire into the opening and tighten the screw. I use fairly big

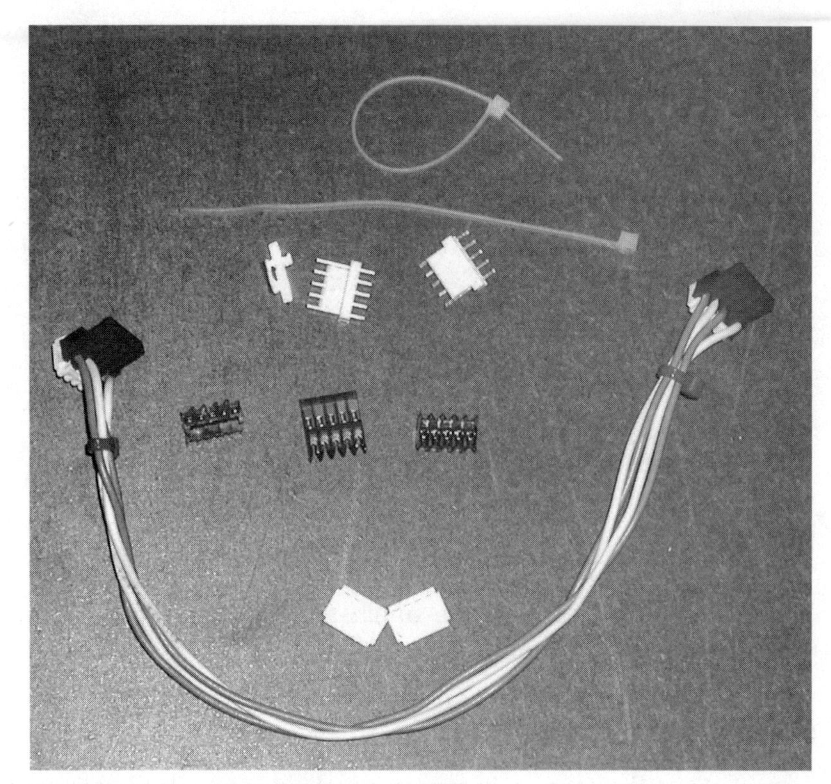

Figure 2-12 The MTA connectors from AMP.

Figure 2-13 Basic barrier strips.

Figure 2-14 Screw connectors are very easy to use.

ones (available at Radio Shack). The devices I have from BDMicro all have much smaller screw connectors, but they work the same way.

Summary

- Electronics is pretty much just the application of electrical laws.
- Electronic devices are a fancy way of converting electrical power to heat.
- Most of the time I spent on electronics was in making cables and connectors.

Chapter

3

Internal Communication: The Control Network

For a robot to function smoothly (or at all), there has to be a way for the main processor to communicate with the various peripherals (sensors, motors, and that sort of thing). Sometimes this is trivial and sometimes it's a pain, but it has to be done. If the robot cannot control the motor speed, it's not going anywhere (well, due to a mistake I made, I did have a robot that was temporarily able to start the motors without stopping them, but this was easily fixed when I finally figured out what I'd done wrong).

Allow me to interject a few definitions here. Networking protocols often are described by the roles played by the participants:

- A *node* is a "thing" that communicates on the network, such as your Linux motherboard or a motor controller. There may be rules governing when and why a specific node may communicate.
  - A *master* is a node that can control the network; some networks may have more than one master. Your Linux motherboard will be a master.
  - A *slave* is a node that can communicate only when requested to do so by a master. A motor controller might be an example of a slave node.
- *Peer-to-peer communication* happens between equals (Figure 3-1). Ethernet is an example of a peer-to-peer network. Any

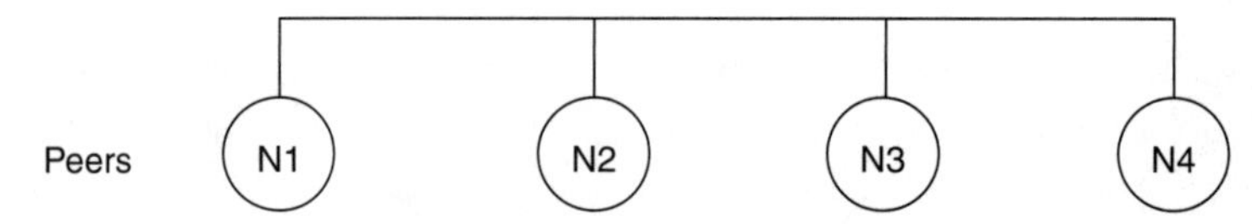

Figure 3-1 Peer-to-peer communication.

node can initiate communication; therefore, all the nodes are equals. In a peer-to-peer network there must be some way of handling the case in which more than one node tries to communicate simultaneously.

- *Master-slave communication* happens where one node is the master and the other nodes are slaves (Figure 3-2). All communications are initiated by the master node. This is not meant in any sort of pejorative sense.
- *Multimaster communication* is just like master-slave communication except that more than one node can be a master node (Figure 3-3). Multimaster networks have the same problem that peer-to-peer networks have.

Also, there are a couple of definitions I use:

- A *module* is a single controller for one or more items. Typically, I buy or make the modules.
- A *pod* is an enclosed collection of one or more modules. The pod communicates on the network via a serial protocol such as MojoBus (discussed later) or Ethernet. I typically make the pod from a set of modules.

This makes it easy for me to create different units that can be plug-and-play. It also allows me to mix items from different manufacturers to create a pod. For example, I can create a distance-sensing pod by combining a RoboBRiX sonar board that also can control a servo (by panning), three Sharp IR Rangers,

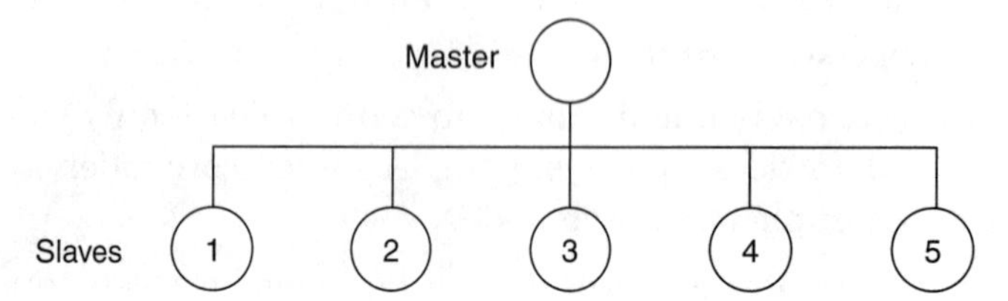

Figure 3-2 Master-slave communication.

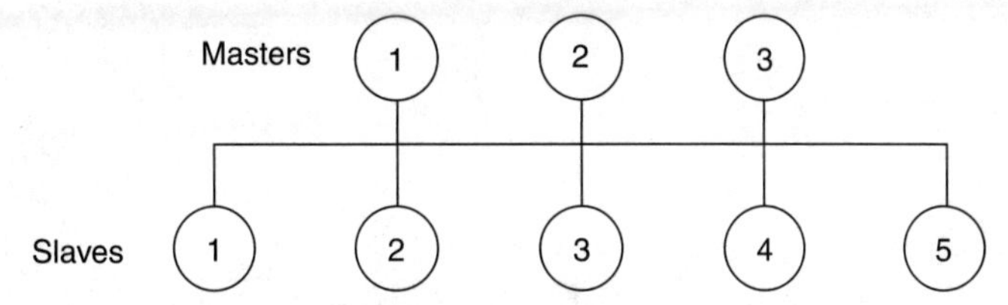

Figure 3-3 Multimaster communication.

a microcontroller chip to handle the IR devices, and MojoBus communications (Figure 3-4). I put the whole group into a box I bought at Radio Shack, and poof: I had a pod.

I also could have a pod that consists of a dual motor controller combined with multiple sensors to read odometry information and adjust the speed of the motors accordingly (Figure 3-5).

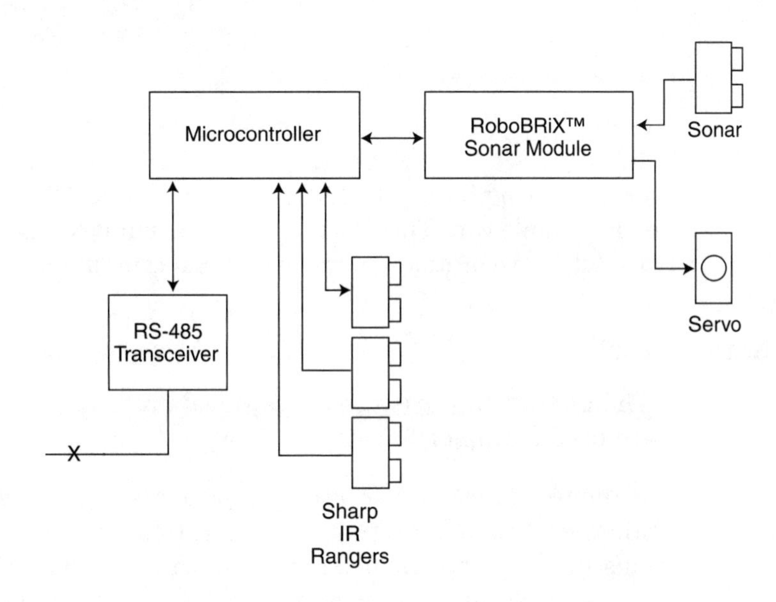

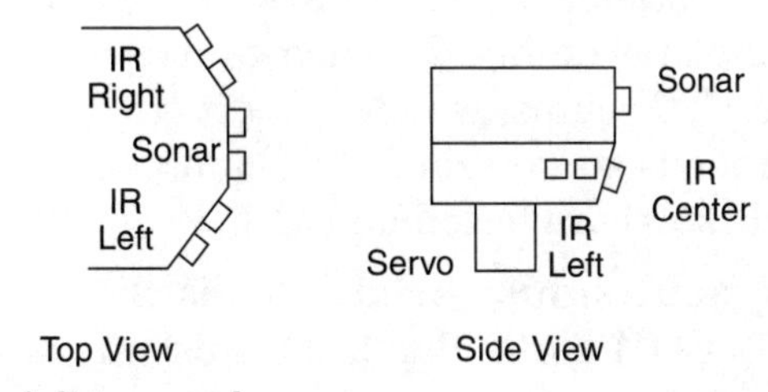

Figure 3-4 A distance pod.

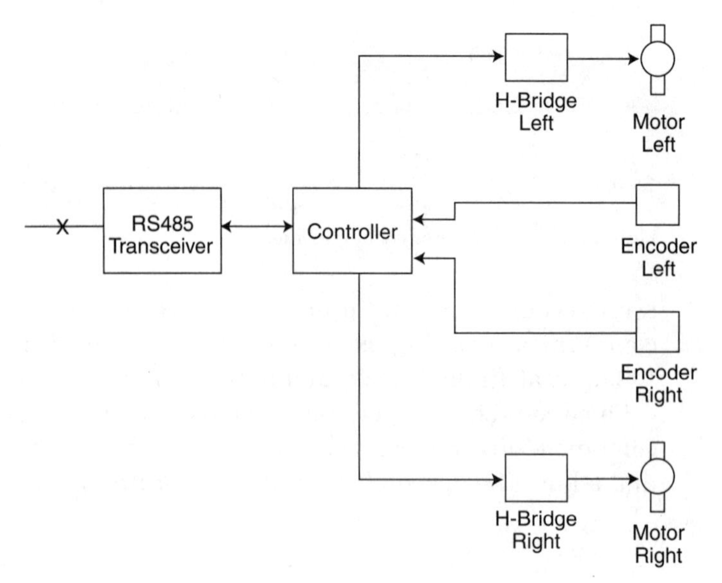

Figure 3-5 A motor pod.

I am working on a pod that can control up to 128 LEDs in a programmable way. This is a trivial device, but it will be extremely useful for decoration and for nonverbal communication.

Hardware Layer

The physical layer is the lowest-level protocol and the hardware used to implement it.

- **Ethernet.** Since almost all motherboards now have built-in Ethernet, this is tempting. However, there are a lot of problems in using this for most internal communication. I prefer to reserve Ethernet for communicating with other computers.
- **USB.** This isn't a bad choice. USB 2.0 is extremely fast, and many standard drivers are built into the operating system. You can even build your own devices by using chips such as the FTDI chips, which convert from USB to your basic asynchronous serial protocol. A lot of inexpensive devices (such as webcams) are based on USB.
- **Asynchronous serial.** This is the basic protocol used by a PC's COM ports. Basically it uses 7 or 8 bits per byte, 0 or 1 parity bit, and 0 to 2 stop bits to communicate a byte. Since

there is no clock line, the protocol assumes that each side knows what speed the other is using. It is possible to design a program to detect automatically what speed (or baud rate) the other side is using. That is not the subject of this book.

- **RS232.** A PC's COM ports use RS232. The most important thing to remember is that RS232 is only an electrical standard. It defines what voltages are 0 and 1, but it doesn't define how the 1s and 0s are used. One problem with RS232 is that absolute voltage definitions are used to define a 0 or a 1. This makes RS232 more vulnerable to electrical noise.
- **RS485.** Again, this is an official electrical standard. The biggest advantage over RS232 is that RS485 uses differential communication: There is a twisted pair of wires for each direction, and the difference between the values determines the bit value. RS484 normally is used to communicate in one direction at a time.
- **TTL-level.** This is just like RS232, but it works directly at the electrical level used by the integrated circuits. Typically the signal is reversed, with a high value indicating a 0 and a low value indicating a 1. (Why? I don't know, and don't care; I just use what I have available.)

- **I2C.** This is built into most PC motherboards, primarily for use with temperature sensors. I2C stands for Inter-IC Communications and is meant to be used for short distances. Unfortunately, my iBase 890C motherboard doesn't have a documented I2C connector. Typically, this is called the SMBus connector.
- **SPI.** The serial peripheral interface is nice for short runs on a circuit board, but I wouldn't want to use it for long-distance communication. It is fast, synchronous, and has the advantage that the master and the slave are both sending data at the same time. The major disadvantage is that a slave-select line must be used for each slave.

Each of these items has a place in robotic communications, some more than others.

Ethernet

Since Ethernet is everywhere, why don't I recommend it for the internal communications of your robot?

Ethernet has several advantages and only one disadvantage. The advantages are that:

- Ethernet is fast.
- Traditionally Ethernet really means TCP/IP over Ethernet. TCP/IP is capable of error correction and guaranteed delivery.
- Each device is addressed easily.
- Ethernet is fairly resistant to electrical noise.
- Many of the hard details are taken care of in the hardware.
- Your motherboard probably has at least one Ethernet port.

The disadvantage is that:

- It's expensive to use. Although it may be built into your motherboard, the peripheral processors used for controlling sensors and motors generally don't have this at a reasonable price. I would estimate that it costs about $50 per peripheral to make it Ethernet-enabled. However, this may change in the future. Perhaps by the time you are reading this book, it will have changed.

USB

The Universal Serial Bus (USB) is becoming the most commonly available serial protocol used on PC motherboards.

The most common use of USB on a robot is to connect the motherboard to a camera. Keyboards, mouses, and hard drives also may be connected. In addition, you can connect sensors via the USB port. One company produces Phidgets, which are USB sensors and controllers. Although they are meant for the Windows platform, there are drivers that work under Linux. I will be using some Phidgets in my examples.

Another company (FTDI) produces chips that convert the standard asynchronous serial to USB. The advantage to this is that a standard serial driver that is built into the Linux kernel can be used for this chip. I use a commercial product containing this chip to convert USB to RS485; it can run at almost a megabit per second. FTDI is only one of the companies that produce these chips.

Some microcontrollers can handle the USB at a hardware level. However, this isn't common yet.

The big advantage of an USB is that it is fast and very easy to use. The disadvantage is that the USB protocol is fairly complex. There are a few books (I like *USB Complete* by Jan Axelson, 3rd ed., Lakeview Research, Madison, WI, 2001) and some information on the Web about the inner workings. In a later chapter, I will show a complete USB driver for a Phidget device.

Asynchronous Serial Protocol

This is one of the oldest protocols used on computer motherboards. It is the same protocol used by the serial ports of a motherboard. However, various means may be used to send the bits (RS232, RS484, smoke signals, etc.); that means there are some different rules.

For the most part, however, this is fairly simple:

1. One start bit (1).
2. Seven or 8 data bits (a combination of 1s and 0s).
3. An optional parity bit:
   a. No parity.
   b. Odd parity: This bit is 1 if the total of all data bits and the parity bit is odd.
   c. Even parity: This bit is 1 if the total of all data bits and the parity bit is even.
   d. Mark parity: This bit is always 1.
   e. Space parity: This bit is always 0.
4. One or more stop bit(s).

The normal standard for communication is 8N1, which means 8 data bits, no parity bit, and 1 stop bit.

I use the RXTX package (Gentoo package rxtx) for serial communication. This package uses the same API as the javax.comm package but was created under the GPL license. Because of licensing issues, the package is gnu.io instead. The RXTX package works under both Linux and Windows. You can download the package from http://www.rxtx.org.

The code that I use to create the serial connection I use for Groucho is as follows:

```
InputStream in = null;
OutputStream out = null;
```

```
// Create the connection
RXTXCommDriver driver = new RXTXCommDriver();
serialPort = (SerialPort) driver.getCommPort("/dev/tts/USB0",
                                  CommPortIdentifier.PORT_SERIAL);
try {
   serialPort.setSerialPortParams(115200, SerialPort.DATABITS_8,
                      .STOPBITS_1, SerialPort.PARITY_NONE);
   in = serialPort.getInputStream();
   out = serialPort.getOutputStream();
} catch (Exception ex) {
   ex.printStackTrace();
}
```

Note that I'm using all of this code merely to get the input and output streams for transferring data. I then pass the streams to the constructor of another object.

RS232

The RS232 standard uses 12 V for a 1 bit and 0 V for a 0 bit. The advantage of using this standard is that many devices already use it. The disadvantage is that it easily can be affected by electrical noise.

Another disadvantage is that many laptops do not handle the signal levels of the RS232 standards properly.

RS485

Like RS232, RS485 is an electrical standard. This standard says nothing about how data are transferred except as bits. The physical wiring consists of twisted pairs (one pair per direction) plus a ground wire.

RS485 typically is used in half-duplex mode. This makes some things easier and some things more complex.

Basically, each transmission line (send, receive, or both) uses a twisted pair of wires and the value of the bit depends on the difference between the voltages. There also should be a ground and a power line, but they can be separate. This twisted pair minimizes distortion and the effects of other electrical systems. The measurement of a difference as opposed to an absolute voltage also makes RS485 much more robust than RS232. Also, using a twisted pair of wires for a single communication direction means that any electrical distortion that happens to one of the wires probably will happen to the other wire in the pair.

You can use a variety of wiring and connectors to communicate RS485. I like to use basic twisted-pair wires connected via standard wire connectors. I've seen RS484 run over phone wire, CAT5 cable, and normal serial cables. Frankly, if I could find some easy-to-use RJ45 (Ethernet) sockets, I might stick with CAT5.

I2C

Intra-IC Communications was developed by a major electronics company to allow integrated circuits on the same board to communicate. This is a synchronous protocol that allows multidrop capabilities.

I use I2C (Figure 3-6) to communicate to the Devantech sensors I use (the sonar and the thermopile array, and I also have a pair of compasses I haven't hooked up yet).

Many PC motherboards have an SMBus connection, which is an I2C-compatible bus. Unfortunately, Groucho's motherboard doesn't have this connection.

Many microcontrollers are able to use I2C directly. This is what I use to communicate to Groucho's sonar units (I use an AVR ATMega128 in a MAVRIC-IIB board to mediate between Groucho's motherboard and the Devantech SRF08 sonar units).

SPI

The serial peripheral interface was developed for a similar reason as I2C: to allow integrated circuits to communicate. SPI is

Figure 3-6 The cables between the sonar units are an example of an I2C network.

also a multidrop protocol; however, instead of having addresses in the data packet, there is a chip select line. When this line is driven low (for only one slave device at a time), that device will listen and perhaps respond (simultaneously) to the master. I like SPI for communications within a short distance, but don't like to run it for long distances.

At this time, I'm not using SPI in Groucho, though I have used it in other robots.

ROBIN

The ROBot Independent Network was developed by many people in the Yahoo group "The Robotics Club." I became interested in a previous version of ROBIN while I was writing this book and brought up a second flurry of discussion and simplified the protocol. At the time I am writing this, Brian Dean, the main person behind the protocol, is developing a ROBIN interface for his RX50 motor controllers. I am working on a Java implementation that I use in my robots. Brian wrote a C implementation for the ATMega microcontroller.

This is a very simple master-slave protocol with provisions for multiple masters. Its advantage lies in its simplicity and ease of programming. The standard definition document is only a couple of pages long. I hope this will be adopted as a standard.

The definition for ROBIN can be found at http://www.bdmicro.com/code/robin.

MojoBus

This is another of Brian Dean's protocols. It's a text-based protocol that can be used on RS485 or pretty much any serial line. Although it isn't as efficient as ROBIN, it has the advantage of being easier to debug because it is text-based. The BDMicro RX50 prototype motor controllers I use for Groucho employ this protocol, and so does the MAVRIC-IIB that I use to control the sonar units. I used a simple terminal program to send and receive test data without having to create a complex test setup.

The protocol is fairly simple. It is a master-slave protocol. Each slave has an ID from 1 to 255. 0 is used as a broadcast address to communicate with all slaves simultaneously.

All the commands are case-insensitive. Commands are of the following form:

>>*N* : *COMMAND* [=*X*] ;

Where *N* is the slave address, *COMMAND* is the specific command, and *X* is the optional value to set. Generally, if the "=*X*" is not included, the command returns the value associated with the command. If the "=*X*" is included, a new value is set. A few commands don't have a value and return nothing.

Some common commands: ro indicates that the command only returns a value, rw indicates that a value can be set, and wo indicates a command that cannot returns a value. I have broken the commands into three parts: general commands for all MojoBus devices, commands specific to the RX50, and commands specific to Groucho's sensor controller

General MojoBus commands

- ID (rw) sets the ID—saves the value in EEProm.
- BAUD (rw) sets the baud rate.
- SAVEBAUD (wo) saves the current baud rate in EEProm.
- WHO (ro) returns the full software information.
- ANNC (ro) returns the WHO value delayed proportionally to the ID. This is done so that you can do a ">>0:ANNC;" and see the values sequentially, rather than have all the slaves try to write on the bus simultaneously.

RX50 MojoBus commands

- D (rw) sets the direction and duty cycle of the PWM (values from –100 to 100).
- ENCP (rw) sets the encoding period in ms. I generally use a value of 10.
- KP (rw) sets the P value for the PID calculations (I like 0.01).
- TVEL (rw) sets the target velocity (I used –1000 to 1000).
- AVEL (ro) returns the actual velocity.
- APOS (rw) returns/sets the actual position according to the motor encoders.

Groucho's sensor controller

- SONARS (rw) sets the number of sonars.
- SCAN (wo) starts scanning the sonar units in a round-robin way. Two units on opposite sides are fired simultaneously, and only the first reflection is recorded.
- READ (ro) reads the array of current sonar values.
- STOP (wo) stops the sonars from scanning.
- TPAREAD (ro) reads the array of 9 pixels from the thermopile array.

MojoBus master code

I use the RXTX code to create an RS485 connection and set the various parameters. I use an 8N1 connection with a rate of 115,200 baud.

My MojoBus commands are defined with a MojoCommand class:

```
public class MojoCommand implements MojoConstants {

   public static void main(String[] args) {
   }

   private String command = null;
   private String parameter = null;
   private int destination = 0;

   public MojoCommand(String command, String parameter, int destina-
tion) {
            this.command = command;
            this.parameter = parameter;
            this.destination = destination;
   }

   public MojoCommand(String command, int destination) {
            this(command, null, destination);
   }

   public MojoCommand(String command) {
            this(command, null, MOJO_ADDR_ALL);
```

```
    }

    public String toString() {
        String s = ">>" + destination + ":" + command;
        if (parameter != null) {
            s += "=" + parameter;
        }

        s += ";";

        return s;
    }
}
```

This class allows MojoBus commands to be created and sent via a MojoConnection:

```
public class MojoConnection {
    private BufferedReader reader = null;
    private BufferedWriter writer = null;

    private boolean lock = false;

    public MojoConnection(InputStream in, OutputStream out) {
        reader = new BufferedReader(new InputStreamReader(in));
        writer = new BufferedWriter(new OutputStreamWriter(out));

        try {
            while (reader.ready()) {
                int i = reader.read();
            }
        }
        catch (Exception ex) {
            ex.printStackTrace();
        }
    }

    public String readLine() throws IOException {
        String s = reader.readLine();
        while ((s == null) || (s.trim().length() == 0)) {
            s = reader.readLine();
        }
```

```
            return s;
    }

    public void writeLine(String line) throws IOException {
            writer.write(line);
            writer.newLine();
            writer.flush();
    }

    public boolean ready() throws IOException  {
            return reader.ready();
    }

    public synchronized boolean getLock() {
            if (lock == false) {
                    lock = true;
                    return true;
            }

            return false;
    }

    public void writeCommand(MojoCommand command) throws
IOException {
            writeLine(command.toString());
    }

    public void waitForLock() {
            while (!getLock()) {
                    Thread.yield();
            }
    }

    public synchronized void releaseLock() {
            lock = false;
    }
}
```

You'll notice that there is a voluntary locking mechanism. This must be used when the MojoConnection is used. If it isn't used, there is a possibility of having two threads use the connection simultaneously. The lock should be kept until any response is received.

An example of use from the MojoMotor class is

```
connection.waitForLock();
MojoCommand command = new MojoCommand("TVEL", "" + p, mojoId);
try {
    connection.writeCommand(command);
} catch (Exception ex) {
    ex.printStackTrace();
}
connection.releaseLock();
```

Summary

- There are many possible network connections.
- Networks consist of a physical level and a protocol level.
- You can run almost any protocol over almost any physical network.

Chapter

4

Sensors

This chapter deals specifically with sensors: the devices that measure something in the world and convert it to something readable by a computer (Figure 4-1). To communicate between the sensor and the computer, we will use one or more of the network protocols I wrote about earlier.

One problem with sensors is that the value of a sensor is not always what it should be. Sensors break, some sensors have blind spots, some are more reliable than others, and a dog might be sitting on one. A robot should use a combination of sensors for anything important and be able to work with the occasional sensor failure.

This can't be stressed enough. A sensor returns some numeric data that the computer must resolve into a measurement of the world, not the state of the world itself.

There are several ways to connect sensors to a computer.

1. **Direct connection to the PC motherboard via a standard protocol.** This can be done via the SMBus port or the RS232 port or even the USB port. Some sensors can communicate via I2C on this bus, which allows you to use those sensors without an intermediate controller. An example of such a sensor is the Deventech SRF08. Some motherboards also contain some direct digital I/O. The iBase 890c motherboard I use for Groucho is one of those boards (unfortunately, it doesn't contain the SMBus port).

   The RS232 port, generally called the serial port, is available on most motherboards. Unfortunately, it seems to be

Figure 4-1 A sample of sensors.

disappearing. The new Nano-ITX board from VIA doesn't have one, and neither did my last two laptops.

2. **I/O subprocessor.** This is done by connecting one or more I/O processors to the computer that handles the input and output. The main computer communicates with this processor to deal with sensors. An example of this type would be a PicoByte Servio. I built one of my own based on the JStik Java controller. Groucho uses a MAVRIC-IIB board. This is a fairly traditional way of doing things.
3. **Smart sensors.** This is done by connecting single sensors and processors to the computer via one or more ports, such as the serial port or a USB port. An example of this type would be Phidgets USB devices such as the PhidgetAccelerometer. Actually, a PhidgetInterfaceKit 8/8/8 can handle eight digital inputs, eight digital outputs, and eight analog inputs. Thus, it behaves more like an I/O subpocessor, but the other sensory Phidgets fit this category.

The advantage of putting sensors on a bus is that it is easy to add new sensors. The disadvantage is that all the sensors have the same bandwidth.

Although I will give an example of the first type, most of the examples will use the latter types.

There are several ways to categorize sensors:

1. **Internal/external.** An internal sensor measures the state of the robot, and an external sensor measures the outside world. Sometimes the categories blur.
2. **Passive/active.** Passive sensors merely read something from the environment; active sensors send out a signal to modify the environment before taking the reading.
3. **Type of measurement.** Sensors can measure distance, time, force, magnetic fields, and intensity of light.
4. **Type of output.** Sensors can communicate their values by changing resistance, changing voltage, turning on or off, using a protocol such as SPI or I2C, or changing the width of a pulse. I call sensors that merely change a value, such as resistance or voltage, basic sensors; this includes sensors that turn on or off.
5. **Simple/smart.** Simple sensors return a value (usually converted to a number by an I/O processor of some sort) to the computer, which then interprets it. Smart sensors can process the data first so that the computer doesn't have to do as much work. An example of a simple sensor is one that returns a voltage corresponding to a distance; an example of a smart sensor is one that detects doorways.
6. **Contact/noncontact.** A bumper switch and a whisker (a lightweight bumper switch) are examples of contact sensors. They have to have contact with something to sense it. A sonar unit is noncontact in that the sensor itself doesn't make contact with a physical object.

Connecting Sensors

Since Linux is not a real-time operating system and the Mini-ITX hardware I have doesn't include sensor-friendly hardware, such as analog-to-digital converters (ADCs), you have to go with an I/O processor of some sort. Sometimes you can deal with a binary (on/off) sensor via the parallel port or the serial port, but I wouldn't count on having these parts in future computers. My newest laptop has only USB connectors, and the new VIA Nano-ITX boards have dropped the serial port also. However, if you want a serial port, there are USB-to-RS232 cables. I suggest that you get the ones with the FTDI chipset because they seem to be supported better by Linux.

The separate I/O processor

An I/O processor is a device, usually based on a microcontroller, that has several different I/O options.

The MAVRIC-IIB (Figure 4-2) has multiple analog inputs, servo controllers, and several ports that can do either digital output or digital input. More important, it can talk I2C to my sonar units. I use the MojoBus protocol over RS485 to communicate with this board. Although the base code was written by Brian Dean, I modified and added to this code to have the board do what I wanted it to do. This code will be posted on the Web. I connect the Devantech sonar units and the Devantech thermopile array unit with this board.

Figure 4-2 The microcontroller I use as a sensor controller.

The prototype RX50 motor controllers use the same MojoBus protocol. Although they control the motors, they are also sensors in that they read the motor encoders. Because of their reading the encoder values, the approximate distance can be obtained. This isn't a general sensor controller, but it works for the specific needs of a motor controller.

An alternative is the Phidget Interface Kit 8/8/8. This is a USB device that has eight analog inputs, eight digital inputs, and eight digital outputs. Right now, I'm only using one analog input to handle the dog-toy sensor and two digital inputs to handle the side whisker switches.

As an alternative, I constructed an I/O processor of my own with a Systronix JStik (a nice Java processor) and an I/O board that I designed (Figure 4-3). Mainly, I used the JStik as the

Figure 4-3 My home-brewed sensor processor.

main processor for a robot. I haven't used this system for a while.

A commercial alternative is the Picobyte Servio. It has eight analog inputs and can control more servos than I can image. To communicate with the Servio, you merely send command bytes over the serial port and then read the responses. Since my laptop has only a USB port, I use a USB–Serial cable for testing. It works wonderfully.

No matter what sensor controller is used, the basic principle is the same. Sensors are connected to the I/O processor. Depending on the processor, you may have to make your own cables, but this becomes easy with time and practice

Smart sensors

If you want an off-the-shelf solution, Phidgets are a great choice. These are devices that connect via the USB port and can function as sensors, as I/O controllers, and as motor controllers, depending on the Phidget chosen. Although they cost more than making your own sensors, they are a very good solution and require very little work and time to start.

Unfortunately, they are fairly slow as sensors go. Since they use the Human Interface Device (HID), which was meant for such things as mouses and keyboards), USB drivers don't report data as quickly as could be wished. However, they are fast enough for most needs.

The advantage of using the built-in USB ports can overcome the disadvantages of lower speed. I find that the speed is good enough.

The disadvantage is that each one takes up a USB port. Although Groucho's motherboard has six USB ports, those ports get used up quite quickly.

RS485

I find that RS485 is better for data communication in a robot than RS232 is. Since most PC boards don't contain RS485 ports, you may need an RS232 or USB-to-RS485 converter. These converters can be made easily by anybody with a little experience in electronics, or you can buy them from a number of sources. I bought mine; it converts between USB and RS485.

The main advantage of RS485 is the noise immunity of the differential voltage and the twisted pair cables.

Modules and pods

To make things easier on the main computer, I could organize the sensors into modules that can be combined into what I call "BotPods" or just pods. Each pod has an RS485 connection that uses a common protocol such as MojoBus.

One advantage of combining sensors into pods is that a pod can be designed to preprocess the raw data from the sensors and put less load on the network and the main computer. For example, I can make a pod with two IR distance sensors and a sonar range finder. Not only can I preprocess the data to return the distance in inches (or centimeters, millifurlongs, or whatever), but by careful placement of the individual sensors, I can have the pod return the clearest path.

Hardware limitations sometimes limit the choice of baud rate to 115k. This is only 10,000 characters per second, and the sensor subsystem must be thought out carefully to maximize the use of the bandwidth. This is one of the reasons I prefer to use smarter sensors.

Frankly, there isn't any speed fast enough to communicate all the data that you will want over an inexpensive data path. And, when the network gets faster, you'll want more data faster. The key is to find a network that works fast enough for the sensors you use.

Example Sensors

The bots in this book use several kinds of sensors. They should span the range of sensors and give you an idea of what kind of data you can expect. I will also show some simple circuits that use these sensors.

The switch

A simple on-off switch can be used to gather a lot of data about the world. These sensors are extremely inexpensive and can be read with a digital input (Figure 4-4).

The most basic use of a switch is as a bumper switch or whisker switch. That is, the switch is physically turned on (or

Figure 4-4 Some basic switch circuits.

off) when the robot hits an obstacle. This works fine for small robots and as an emergency signal for larger robots (larger robots have more momentum and, by the time they actually touch an object, stopping may not be an option).

The main difference between a bumper and a whisker is that a bumper is bigger and a whisker is smaller and should not damage walls and such.

However, you can use a switch for other sensory tasks:

1. **Jolt sensing.** Sometimes you want to know if the robot is running over rough terrain or has hit something. If you mount a switch below the switch, it will turn on and off if the robot moves up or down.
2. **Tilt sensing.** Sometimes you want to know if a robot is tilted or is accelerating in one direction. By using a pendulum, you can turn the switch on or off, depending on the tilt or acceleration (Figure 4-6). This is very similar to jolt sensing.
3. **Human input.** Sometimes you want to have a simple switch that will do something boring such as turning a robot on or off. Or, perhaps the switch puts the robot into a different mode.

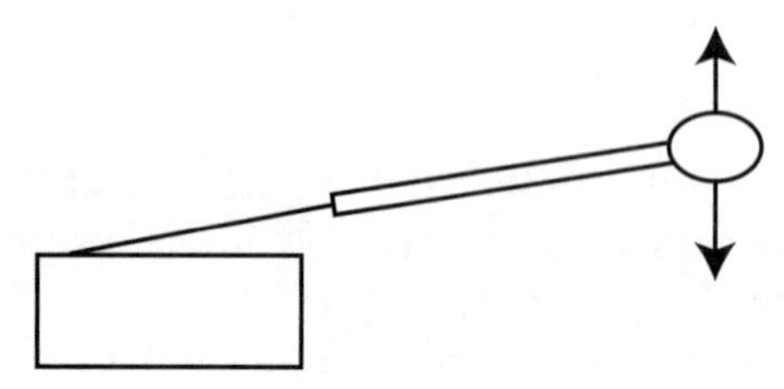

Figure 4-5 Jolt switch.

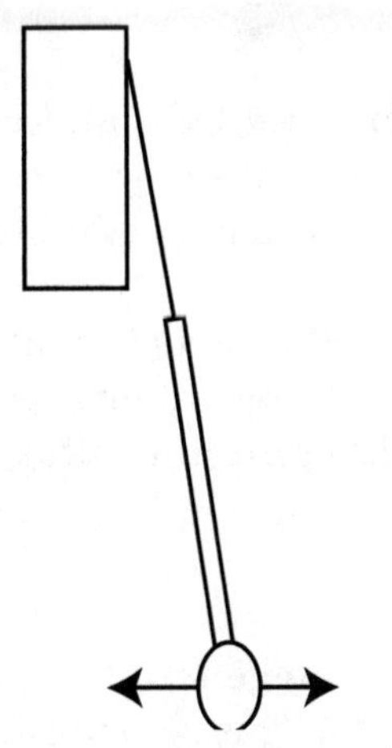

Figure 4-6 A tilt switch.

The Photoresister (CdS cell, photocell, or light-sensitive resistor)

The cadmium sulfur cell is an inexpensive (under $3) light sensor. It changes resistance as the amount of light changes. This is easy to interface to any analog input. Another advantage is that the CdS cell's sensitivity to color is similar to that of a human eye.

As with any sensor, you can read a value from the cell at any given time. From this information you can get more information than you might think.

1. By reading the current value, you can tell the light value at a specific time.
2. By reading the value over time, you can differentiate these values by estimating the slope of the curve. If you do this, you can tell the rate of change of the light and know if the light is increasing or decreasing.
3. By using two CdS cells and calculating the difference between the cells, you can point the robot at the brightest (or darkest) area.
4. By reading the values of two CdS cells over time, you can tell in which direction the change of light is going.

Infrared detection

By using a device that senses infrared light and an IR transmitter (an IR photodiode or phototransistor), you can sense obstacles or changes in color. There are many different kinds of sensors that use IR.

The most common type is one that detects the intensity of IR that is returned. The problem with this scheme is that different colored objects may return different intensities.

Passive infrared detector (PID)

This is a sensor that detects the change in infrared light. It is used commonly in motion detectors. This sensor can be used to distinguish moving animals and people from the background. Or, the sensor can be moved. The typical sensor gives a single pulse when the temperature (the IR radiation reaching the PID) changes significantly. I have some that will return a pulse that is about a second long.

The sharp IR ranger

This is an interesting device that can tell the distance from an object. In smaller bots, I use this detector as my main (or only) means of range detection.

These sensors work by sending out a focused beam of infrared light from one point and receiving the reflection from another (Figure 4-7). The key is that the angle of reflection is measured with a position-sensitive sensor.

There are three varieties of these, with different ranges from 30 to 150 cm.

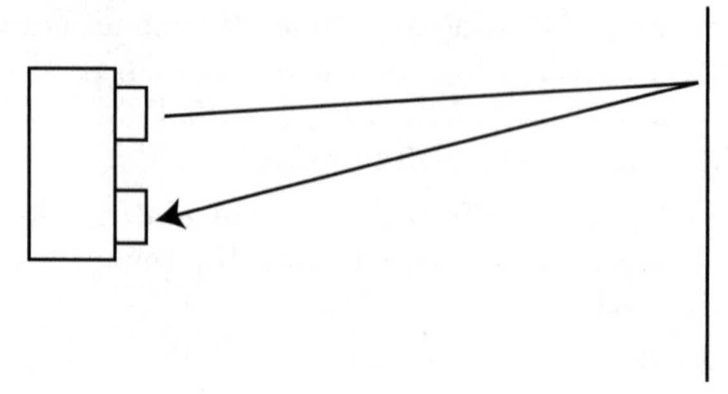

Figure 4-7 A ranger working.

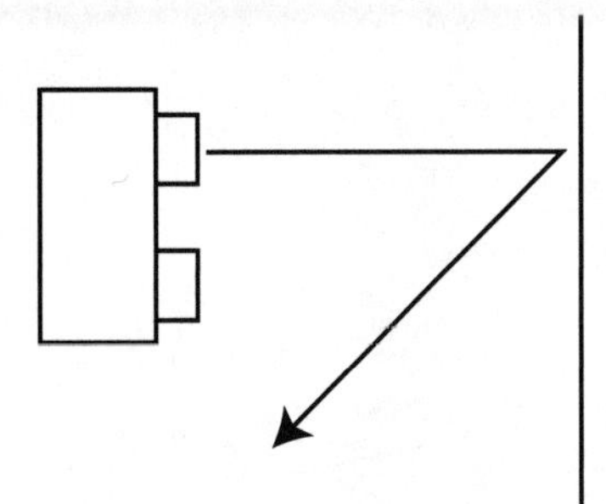

Figure 4-8 A ranger too close.

If an object is detected at the minimum sensing distance, the voltage produced is at its highest. If there is no object in range, the voltage is at its minimum.

One minor disadvantage of this sensor is that if an object is less than the minimal sensing distance, a low voltage reading will be produced (Figure 4-8).

I used one of these as the front dog toy detector. I found I needed to add it because Groucho's front caster could be blocked fairly easily.

Sonar

The sonar sensors are also range detectors, but they can operate out to a longer distance. The ones I'm using are rated at 3 inches to 10 feet. The basics of sonar are very simple: You send out a pulse and then measure the time it takes for the pulse to come out (Figure 4-9).

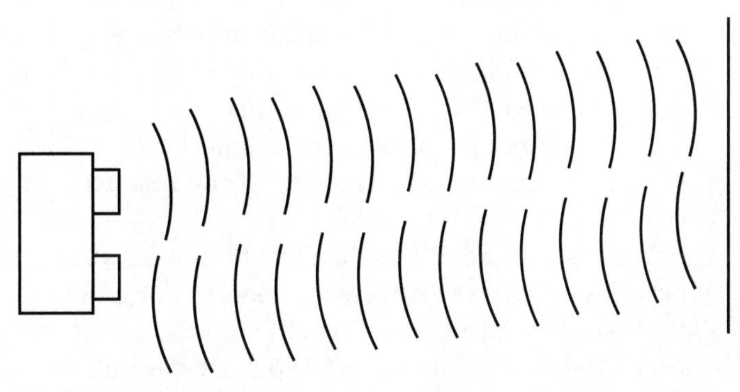

Figure 4-9 A sonar pulse.

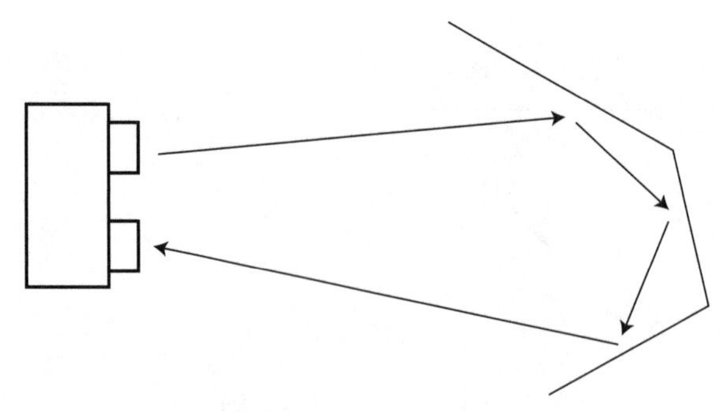

Figure 4-10 A confused sonar.

The downside to using sonar is that certain configurations of obstacles can confuse the sonar by delaying the reflection (Figure 4-10).

I have been using Devantech sonar units. The basic one, the SRF04 is very simple, and you need to use a microcontroller to deal with the timing issues. Later models, such as the SRF08 I use, have microcontrollers installed. They use I2C to communicate and can be connected directly to the motherboard's SMBus (if your motherboard has one).

Accelerometers

Accelerometers measure the force of acceleration. This can happen by traditional acceleration or by tilt (tilt measures the acceleration of gravity). By using a MEMS device, they can produce an accelerometer the size of a normal IC, using extremely low power.

The accelerometer I use is the Phidget Accelerometer, which handles all the low-level stuff in hardware and communicates via USB. I did have to write a low-level driver for this, but USB drivers are easy to write for Linux.

A few years ago, these accelerometers were expensive. Now, they are very inexpensive since they are used in the automotive industry.

You have to do a bit of coding to use an accelerometer properly, because of the effects of gravity. For example, if the robot is resting on an upslope, it will sense a net acceleration to the rear (and down, but I don't typically care about up and down) (Figure 4-11).

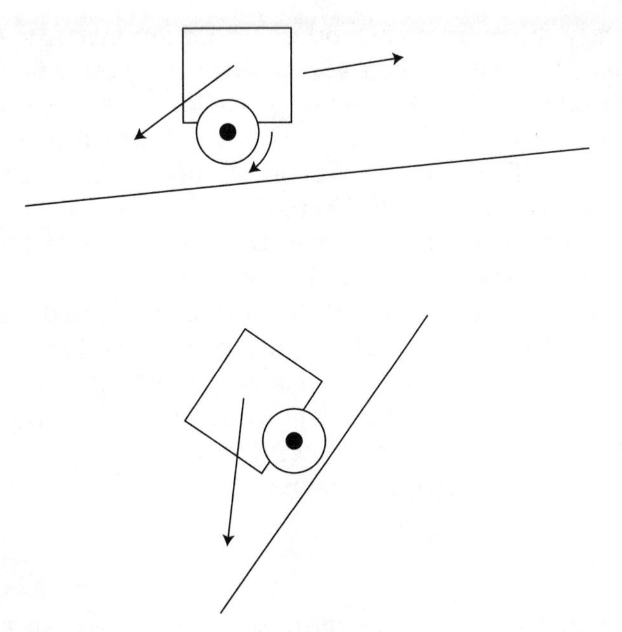

Figure 4-11 The robot accelerating and at rest on a slope.

Gyroscopes

Also a MEMS device, the gyroscope I use is a yaw-rate gyroscope. It doesn't measure an absolute position but reports the current rate when requested. Through integration, you theoretically can get the absolute direction. In practice, it doesn't work as well as it does in theory. However, knowing this, you can still use the values and correct the heading with other sensors when possible.

I have not added this to a robot yet, though. It's sitting in its box.

Compass

A compass, or magnetometer, is one way to find an absolute direction in the world. Well, not quite absolutely. Unfortunately a magnetometer works by sensing the magnetic fields around it. This means that even the robot's motors can throw the compass off. The birdcage also causes problems.

Capacitance sensors

Capacitance sensors detect a small change in capacitance as objects in the world change distance from two oscillators that slightly differ from each other. The difference frequency (also called the beat frequency) changes as the outside objects add themselves to the equation. This effect was discovered in the early 1900s and was the basis of the musical instrument the theremin. The version I'm using is called Thereminvision. Like the accelerometers, the Thereminvision unit produces a series of pulses where the length of the pulse reflects the beat frequency. The base unit has four sensors.

This is a nice sensor in that there really aren't any downsides. Although it takes a bit of work to calibrate the device, once you've done this, it seems to work well.

The range is variable, depending on the antennas and the size of the object to be sensed. The sensitivity depends on the precision of the pulse measurement.

Odometers

Odometers are devices used to measure distance traveled. Usually they measure how much a wheel or something connected to it rotates. This is not the same thing as actual distance traveled. Wheels can slip; robots can bounce. However, coupled with other sensors, odometers can be very useful.

Odometers come in two varieties: relative and absolute. Relative odometers send signals at intervals as they turn. Absolute odometers send a direct value showing where the wheel is. You can make your own encoder for an odometer, but that is beyond the scope of this book. I think that commercial encoders are inexpensive enough to use.

The motors I use have encoders already installed. Unfortunately, the encoders are attached directly to the motor, and the motor is geared down a lot. This means that one rotation of the wheel produces over 800,000 pulses. Luckily for me, the RX50 H-bridges I use can handle this speed. It's also lucky that I know the creator, Brian Dean. He helped debug my motor problems and rewrote the firmware several times.

Code

The Java code for sensors is some of the more convoluted code in my framework.

It includes the following:

- **Sensor:** The interface to a generic sensor.
- **AbstractSensor:** The common implementation of a sensor.
- **SensorImpl:** The interface to the hardware. Every sensor instance has a SensorImpl to actually read the sensor.
- **AbstractSensorImpl:** The common implementation of a SensorImpl.
- **SensorThread:** The thread that runs in the background and reads the sensors at intervals determined by the sensors.

In addition, a few types of predefined classes extend AbstractSensor and AbstractSensorImpl.

Groucho's Sensor Architecture

Groucho has a reasonable number of sensors, discounting audio and video (Figure 4-12):

- A sonar ring of 12 sonar units mounted on the bottom of the third deck.
- A Sharp IR Ranger in the bottom front designed to keep the front caster from hanging up on a dog toy.
- Two whisker switches on the sides.

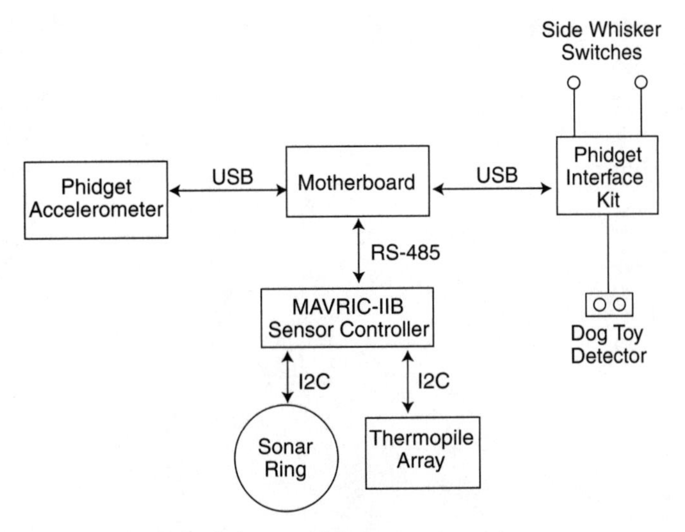

Figure 4-12 Groucho's sensor block.

- A thermopile array in the front. This is designed to detect heat sources, such as dogs and children.
- Two voltage detectors, so that the batteries can be monitored.

Each of these items has different requirements. The sonar units are controlled by the MAVRIC-IIB in SCAN mode. These units are read by Java 10 times a second. The other sensors are read every 50 milliseconds.

I also have some pseudo-sensors that are used for voice input and remote control over the Web. I'll talk about them in a later chapter.

Summary

- An RS485 network can connect sensors with less noise.
- Sensors are the only way a robot can perceive the outside world.
- Obstacle avoidance and detection is an important use of sensors.
- Sensors also are used for distance measurement.

Chapter

5

Behavioral Programming

To make your robot do interesting things, you can program it in a variety of ways:

- Traditional symbolic artificial intelligence (AI)
- Dedicated programming for your particular situation
- Artificial neural networks (ANNs)
- Behavioral programming

In symbolic AI, the programmers make a model of the world in a symbolic language. This map is compared continually to the world according to the sensors, and the actions of the robot are adjusted accordingly. Unfortunately, this requires rather extensive processing capability and memory. The main trouble is the things that are easy to tell a human (such as “turn left, go right past the sofa”) are more difficult to tell a computer. I’m not even talking about language recognition, but rather the difficulty in programming a computer to recognize a sofa.

With dedicated programming you can easily program a small and specific set of actions for your robot. The dark side of dedicated programming is that your robot will be able to do only this small set of rather specific actions. This can be useful in an industrial setting, where a robot on an assembly line performs the same set of actions over and over again. The robot can even be programmed to notify a human supervisor if something goes wrong.

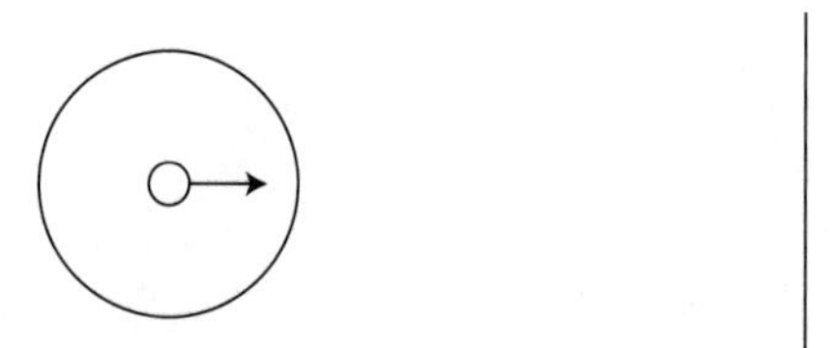

a. The robot goes forward.

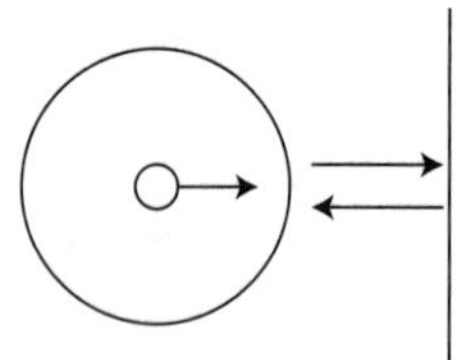

b. The robot senses the proximity of the wall.

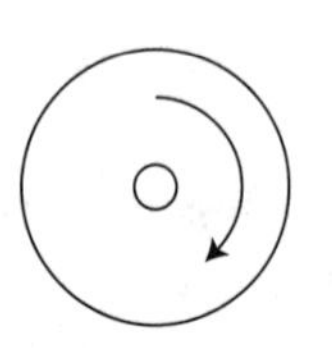

c. The robot turns a random amount.

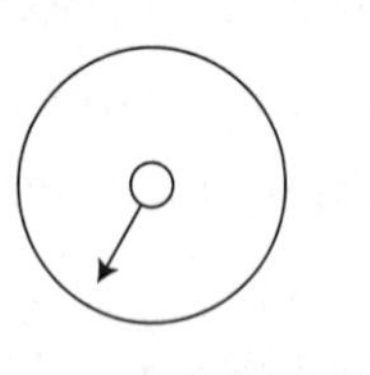

d. The robot goes forward again.

Figure 5-1 A robot with simple behavior.

Artificial neural networks (ANNs) started out as an abstraction of a biological nervous system. However, computer scientists saw the potential and ran with it. An ANN can be used to do anything a traditional program can do. However, a high computational and memory cost usually is associated with an ANN.

Behavioral programming, also known as reactive programming, is a bit of a combination of the above systems: There is no symbolic map of the world, the world as seen through the sensors acts as its own map, and the behaviors are triggered almost directly by the sensors (Figure 5-1). The example robot's change of direction is due to sensing the proximity of the wall.

The important statement above is that the world is its own map. This makes the computation involved in running a robot easier by several orders of magnitude. Yes, your robot may think more like an insect than like a human, but it can appear to react intelligently and in real time. And, in the real world, response time is extremely important.

There is some evidence that our brains use a combination of techniques; we seem to have some maps in our brains that are created with neural networks (real ones, not the artificial stuff), but they are not exact. The input from our senses helps us navigate even in familiar surroundings.

To test this, go into your bedroom at night with the lights off and your eyes closed. It helps to have a sleeping spouse, a messy bedroom, and a couple of dogs underfoot for this exercise. Whenever I try this, even though I've lived in my house for 15 years, I bump into things (give yourself 10 extra points if you avoid waking your spouse). However, with the lights on, I avoid the obstacles (except for a few times when the dogs attempt to evade me and we end up colliding anyway). This means to me that, even though I have some sort of map of my bedroom in my brain (door behind me, bed to the left, dog crate to the front and right, dog probably sleeping at the foot of the bed), it isn't specific enough to navigate without additional information. However, by adding in the sensory perceptions from my eyes, I can navigate easily. In other words, the symbolic map I have of my own bedroom isn't accurate enough for navigation without visual updating (Figure 5-2).

I think this is a sensible combination. Although some things are immovable, many obstacles can change position. Somebody could have moved the dog crate, the dog could have changed where she wanted to sleep, and so on. Some obstacles are immovable: Most of my bedroom furniture is seldom moved.

In the pure behavioral programming flavors, there is no map of the world and little, if any, memory. The behaviors are stateless; that is, they depend only on the current values of the sensors and perhaps a timer. This makes sense from a programming standpoint because it limits the choices available at each point in time.

In the real world, we sometimes want to keep some state and remember bits of the past to make things easier. For example, I may want to trigger a new behavior if Groucho bumps into a wall three times within a short period. However, this can be done without having a map of the world.

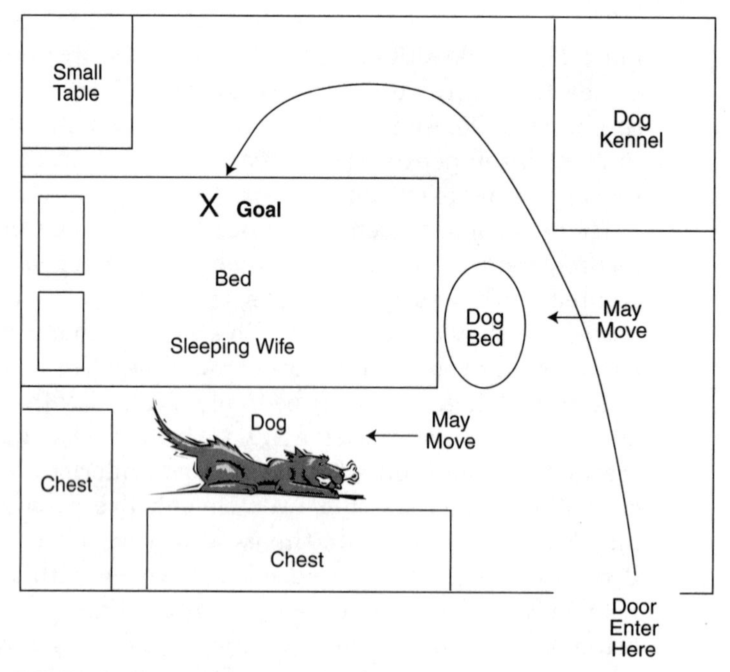

Figure 5-2 Navigating a messy bedroom.

I would like to point out that the term *behavior* in behavioral programming doesn't mean the same as the word *behavior* in referring to human behavior. Human behavior is complex and involves multiple actions, motivations, and occasionally thought. In a robot, *behavior* refers to a reaction or a set of reactions that occurs when triggered by the sensors. I use the word *behavior* for historical reasons. I would rather use the term *reaction*, but I will stick with *behavior*.

I would also like to point out that I think that properly done behavioral programming depends on well-thought-out sensors and their positions on the robot. In other words, if the robot can't sense something, it can't react to it. Therefore, for a robot to be a good candidate for good behavioral programming, the robot needs to be built so that it has good sensor coverage for its mission.

Behavioral Programming History

Behavioral programming is one of the first types of "programming" used by a real robot. In 1953 W. Grey Walter created the

Machina Speculartrix as a thought experiment. He later turned it into physical reality as the Tortoise.

The Tortoise was a simple robot with one light sensor, one bump sensor, two motors, and two vacuum tubes as a “brain.” The drive was a tricycle design, with one motor controlling the direction of the front wheel and the other motor powering the rear wheels. The light sensor was mounted so that it was in line with the front wheel. The bump sensor covered the entire circumference of the robot.

With just these simple things, the Tortoise was capable of many behaviors. It could seek light, hide from light, avoid obstacles, and do other things. The magic was in the combination of simple reactions.

In 1984, Valentino Braitenberg continued this tradition of analog thought experiments in his book *Vehicles*. Using simple sensors connected directly to motors, he created many vehicles, from simple to complex. Even though these are not exactly practical robots, I consider his book an important work for any roboticist to read.

Behavioral programming is not the work of any single scientist but a group effort. Two names stand out in my mind: Rodney Brookes and Ronald C. Arkin.

Rodney Brookes created subsumption architecture, in which high-level behaviors would block or “subsume” lower-level behaviors. His work seems to have mostly been theoretical, though it has inspired a great deal of programming.

Ronald C. Arkin is one of the major researchers in the field of behavior-based robotics (among other things). He has over 120 publications in these fields, including my favorite robotics book, *Behavior-Based Robotics*.

Behavioral Programming Basics

There are many types of behavioral programming, but they share a basic set of characteristics:

- One or more sensors trigger a behavior.
- The active behavior affects one or more actuators (motors) (Figure 5-3).

That’s pretty much it. Notice that there isn’t much “thinking” going on here: The robot just reacts to its environment.

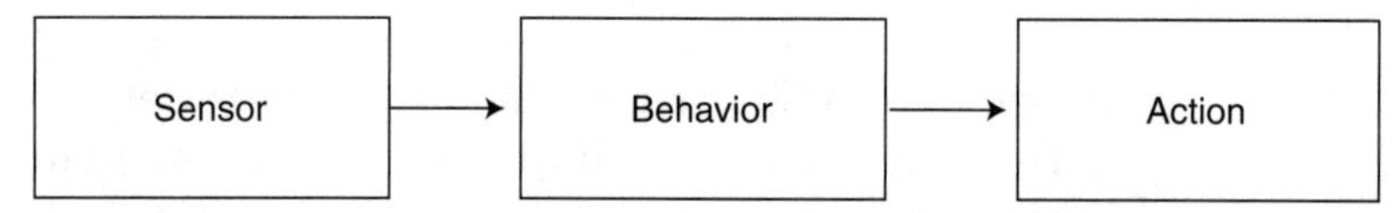

Figure 5-3 Behavior triggering.

However, you can program complex and seemingly intelligent behaviors by using these rules.

As a consequence of this, there is also no learning. The robot has only the behaviors you program into it. However, since you can program a large group of behaviors, this doesn't always matter. There are other hybrid techniques that allow learning.

The main difference between the various types of behavioral programming lies in the way the system arbitrates between different behaviors that are triggered at the same time. Any behavioral system needs an arbitrator, even if it is trivial.

To create a behavioral programming framework, you need a few basic objects in addition to the previously defined Motor, SensorThread, Sensor, SensorListener, and SensorEvent objects (Figure 5-4).

The *Behavior* class contains both a robotic action and an activity level for this action; there will also be a BehaviorListener interface to allow other objects to be notified of changes to a given Behavior. A Behavior instance mediates between one or more sensors and the Arbitrator:

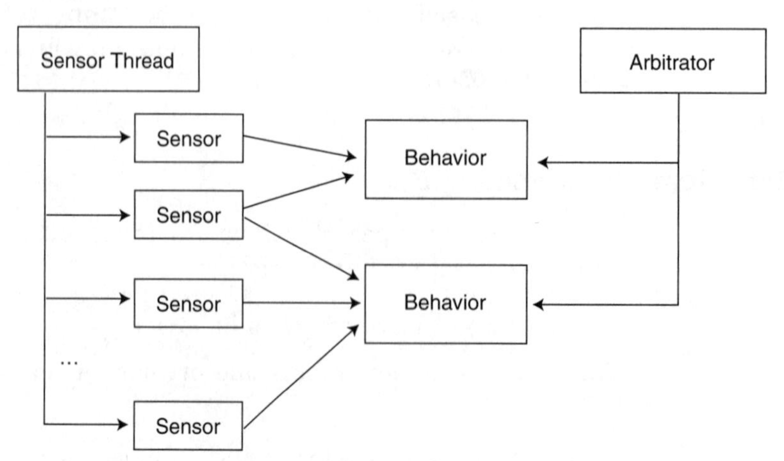

Figure 5-4 The interactions between the classes involved in behavioral programming.

```
Public interface Behavior extends SensorListener {
    // Called once when the behavior is started
    public void start();
    // Called repeatedly while behavior is running; this
    // should not take a long time to complete
    public void run();
    // Called once when the behavior is stopped
    public void stop();
    // Called by the arbitration when the behavior should stop
    public void shouldStop();
    // Returns true when the behavior is stopped
    public boolean isStopped();
    // Returns when this behavior is fully stopped
    public void waitForStop();
    // Forces the behavior to wait before stopping
    public void wait(int millis);
    // Returns an integer corresponding with how much this
    // behavior wants to run at this time
    public int getActivation();
    public void setActivation(int i);
    public void addBehaviorListener(BehaviorListener listener);
    public void removeBehaviorListener(BehaviorListener listener);
}
```

The *Arbitrator* class selects a list of behaviors that want to be active and chooses among them. I have two different arbitrators defined: SimpleSubsumptionArbitrator and the ControlArbitrator.

```
public interface Arbitrator extends BehaviorListener {
    public void addBehavior(Behavior newBehavior);
    public void removeBehavior(Behavior newBehavior);
    public void startArbitrator();
    public void stopBehavior();
    public void startBehavior(Behavior newBehavior);
    public void shouldStop();
    public boolean isStopped();
    public void waitForStop();
    public void chooseBehavior();
    public void chooseBehavior(Behavior behavior);
    public abstract void activityChanged(Behavior b);
}
```

Simple subsumption architecture

Here I will describe a simplified version of Rodney Brooke's subsumption architecture. The full architecture is beyond the scope of this book and merits one of its own.

The very basics of subsumption architecture (Figure 5-5) are the following:

- There is an ordered list of behaviors, with the lower numbered behaviors having higher priority than the later numbered ones.
- Each behavior instance can choose to return either a 0 or a 1 for its activation. If the value is 0, the behavior is not activated.
- Only one behavior is active at any given time.
- The active behavior at any time is the highest-priority behavior that wants to be activated.

The name comes from the fact that the highest-priority behavior that is active will "subsume" the lower-priority behaviors. In its more complete form, subsumption architecture allows behaviors to increase or inhibit sensor readings to

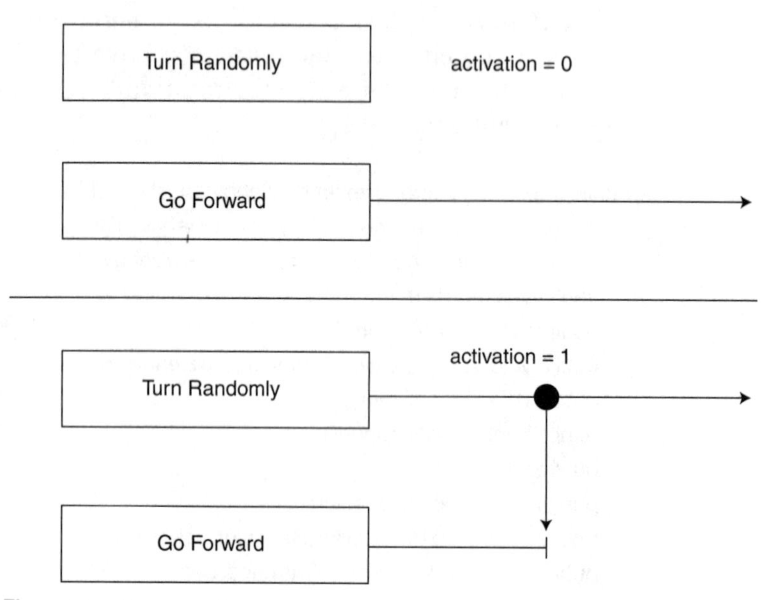

Figure 5-5 Subsumption.

other behaviors in order to follow a biological framework more closely. However, the basics will work for robotics.

To keep the code clean, the Behavior interface implements the SensorListener interface. However, it is possible that the behavior will get information from other sources (different listeners, global variables, pulling it out of the air, etc.).

First, you should make a list of high-level behaviors that you want your robot to do, for example:

- Avoid obstacles.
- Travel in a straight line.

You then need to make a list of the basic actions needed to accomplish those behaviors. This should be a fairly low-level list.

Given Groucho's sensors (a sonar ring and the front short-range detector), we want Groucho to go until an obstacle is close, change direction, and do this again.

If an obstacle is detected, Groucho does the following:

- Change to a random direction.
- Go forward.

Note that to do this I have only two behaviors.

The first one (the one with the highest priority) is activated at the start; this behavior goes forward until an obstacle is detected, and then it stops being active.

The second one (lower priority) always wants to be active but can be active only when all lower-priority behaviors are inactive; this behavior has turn a random amount.

The Code

The actual behavior oriented code in Groucho for this activity is trivial:

```
public void makeBehaviors() {
   // The "go forward" behavior
   // Turn randomly behavior
   Behavior behavior = new AbstractBehavior() {
            public void start() {
                     super.start();
```

```
                        left.setPower(40);
                        right.setPower(40);

                        System.out.println("Started!");
                }
                public void stop() {
                        super.stop();
                        left.setPower(0);
                        right.setPower(0);

                        System.out.println("Stopped!");
                }

                public void sensorChanged(SensorEvent e) {
DistanceSensor sensor =
DistanceSensor)e.getSource();
                        setActivation(
                                (sonars[SONAR_FRONT].isClose() ||
sonars[SONAR_FRONT_LEFT].isClose() ||
sonars[SONAR_FRONT_RIGHT].isClose() ||
                                frontBottom.isClose()) ? 0 : 1
);
                }

                public String toString() {
                        return "forward ho!";
                }
    };
    behavior.setActivation(1);
    sonars[SONAR_FRONT].addSensorListener(behavior);
    sonars[SONAR_FRONT_LEFT].addSensorListener(behavior);
    sonars[SONAR_FRONT_RIGHT].addSensorListener(behavior);
    frontBottom.addSensorListener(behavior);
    arbitrator.addBehavior(behavior);

    // Turning behavior
    behavior = new AbstractBehavior() {
                public void start() {
                        super.start();
                        Random r = new Random();
                        if (r.nextInt(2) == 0) {
                                left.setPower(20);
```

```
                        right.setPower(-20);
                    }
                    else {
                        left.setPower(-20);
                        right.setPower(20);
                    }
                    wait((r.nextInt(10) + 1) * 1000);
                }

                public void stop() {
                    left.setPower(0);
                    right.setPower(0);
                }

                public String toString() {
                    return "turning!";
                }
            };
            behavior.setActivation(1);
            arbitrator.addBehavior(behavior);
        }
```

Notice that I have the two behaviors defined anonymously. I could have created named classes for them, but it was easier to do it on the run.

Limitations

I program my robots by using primarily behavioral programming. However, I admit that there are some limitations to a purely behavioral approach.

These limitations, however, can be compensated for in a hybrid approach in which behavioral programming is combined with some other form of robotic programming.

Learning

Behavioral programming has no provision for learning. A truly autonomous robot should be able to learn from its environment. This is not necessary in all cases, but most of the time it is useful.

Take, for example, a robot that delivers the mail to people in an office. At the very least, it would be useful for the robot to be able to make a map of the office area. This map would allow the delivery robot to take shortcuts and avoid obvious obstacles.

Even in the home environment, it would be nice to have mapping capability if only so that I could tell the robots that certain areas are off limits.

State

Because behavioral programming is based solely on reaction to sensor readings, a purely behavioral system has no concept of state. This means that it can be difficult to differentiate between going straight to go to a goal and going straight to avoid an obstacle.

This can be solved with a few variables or by creating additional behaviors (which adds more complexity to the program).

Timing

Another limitation of pure behavioral programming is the lack of timed behaviors. This is an easy extension, but it takes some thought to add it to a framework.

Timed behaviors are useful when you want to have the robot do something for a specific (or random) amount of time in order to complete its reaction to the situation. For example, when I wanted to have the robot turn to avoid an obstacle, I needed to be able to have the "turning" behavior work until it was finished.

Intuitive programming

Behavioral programming is not always intuitive at the beginning. For example, should the "go straight" behavior or the "avoid obstacle" behavior have a higher priority? This can be a difficult question, because the answer changes with the requirements of the robot.

However, any programming language or environment is confusing at first.

Summary

- Behavioral programming is a strong contender for robotic programming because it requires little memory and runs quickly.
- There are many other types of programming; they tend to be memory- and time-consuming.
- A hybrid approach may be easier and simpler than a purely behavioral approach.

Chapter

6

Audio: Speaking and Listening

One reason to use a powerful computer is that it allows you to add audio communication easily. Both Linux and Windows have applications that can turn text into speech and other applications that can turn speech into text. Not only is audio a powerful means of communication, it adds emotional impact to a robot. I'm sure I'm not the only person who remembers the robot on *Lost in Space* (B9, and yes, there is an active builders' group) saying, "Danger, danger, Will Robinson."

Not all fictional robots can talk. I truly admire filmmakers who can take a robot like R2D2 and give it body language that is easily understandable. However, I'm not doing that for this book. I'd like to, but I don't have a robot capable of decent body language right now. My *next* robot will be capable of body language via a movable head and legs.

In the real world, audio is one of the main ways that we communicate with each other. I can talk to my dog and know that she understands some of what I say (mostly commands and the important words like "food" and "pizza" and "out"). I've taught my parrot to say some interesting phrases over the years. Yet our robots seem to be deaf and dumb.

By speaking, a robot seems more human. When I had just started programming Groucho to speak, I wrote a simple network application so that I could type text and have Groucho "speak" it. I was doing this for a friend's child (who figured out that I was the brains behind this very quickly); my friend asked if Groucho could smile for his daughter. Off the top of my head I replied, "I have no face." The room burst out in laughter.

In other words, having a robot speak can go a long way toward making it more friendly.

Simple text to speech is fairly easy and can be done with an easy to build circuit. However, a more powerful computer can do this better.

Speech to text, otherwise called speech recognition, is more difficult in the general case. This involves not only sensing the sound but recognizing which noise is speech and which noise is, well, noise.

I use two open source packages for audio.

For text to speech I use Festival. Festival is a very powerful speech engine that was developed at the University of Edinburgh. One of the nice features is that it already contains a server so that I can run it as a service in another process. It is available both as a Gentoo package and as source from http://www.cstr.ed.ac.uk/projects/festival.

For speech to text I use Sphinx4, which was developed at CMU. Sphinx comes in three varieties, Sphinx2, Sphinx3, and Sphinx4. Sphinx2 is available as a Gentoo package. Sphinx4 is a purely Java application and can be downloaded easily and built from the Java home site, http://cmusphinx.sourceforge.net/sphinx4. I had to wrap a basic Java server around a modified version of one of the demo programs to have a server in a separate process.

Text to Speech

Festival is easy to set up and can be run as a simple service that can be accessed through a socket. The code to do this is trivial once you have everything set up.

You have to have the proper driver for your sound card; most Mini-ITX boards have the sound chips installed on the motherboard. You can find information about your sound "card" with the *lspci* command, which lists the information on all PCI devices.

For example, Groucho's sound card had the following information:

```
0000:00:1f.5 Multimedia audio controller: Intel Corporation
82801DB/DBL/DBM (ICH4/ICH4-L/ICH4-M) AC'97 Audio Controller (rev 02)
```

This meant that I had to compile the snd-intel8x0 either as a module or to the kernel.

You need to build Advanced Linux Sound Architecture (ALSA) with Open Sound System (OSS), a deprecated sound layer emulation into your kernel.

You need the various ALSA tools. To find out the ones that are useful, I searched the portage tree (*"emerge --search alsa"*).

Importantly, you *need* to turn the volume up! By default, on my system it was at zero, which caused more than a little confusion. I used the application *alsamixer* to fix the volume.

Then, arrange to have Festival started at boot time. In Gentoo you do this with the command *rc-update add festival default*.

Your robot will need a set of speakers. I used a set of Sony nonpowered speakers designed for use with a portable CD player. In my opinion, these speakers don't produce a loud enough voice. In my next hardware upgrade, I will get a pair of powered speakers and see if they work better.

The voices are not the most realistic, but that's all right with me. I like a robot to sound robotic.

Festival is an extremely powerful program. It can be programmed in a dialect of Lisp called Scheme. I tell you this mostly because it is interesting, not because I've used it for anything other than the initialization files and to speak a line of text. This command is *(SayText "Some text goes here")*.

Once you've launched Festival as a service you can just send commands via telnet to (by default) port 1314. This gets old quickly if you're not a great typist. However, you don't need telnet; you only need a way to send text through a socket. This is trivial in most languages. The code to do this in Java is:

```
try {
    // Create the network socket
    socket = new Socket(host, port);
    // Create the writer to send data to festival
    fest_in = new OutputStreamWriter(socket.getOutputStream());

    // Get the stream to get (and ignore) festival output
    fest_out = socket.getInputStream();
} catch (Exception ex) {
    ex.printStackTrace();
}
```

```
if (text != null) {
    fest_in.write("(SayText \"" + text + "\")");
        fest_in.flush();
        text = null;
    }
    while (fest_out.available() > 0) {
    fest_out.read();
}
```

The object fest_in is an OutputStreamWriter that sends text to Festival's input, and fest_out is an InputStream that reads the output from Festival. I found that, if I didn't read the output from Festival, things locked up fairly quickly.

To find out what sort of output Festival has, just do a "telnet localhost 1314." It takes a bit of typing, but you can see the detailed output.

The code above was taken from my robotics framework and simplified a bit. The code in the framework makes sure that the network socket is still open before I write and read to it.

I admit that the speech output from Festival doesn't sound quite like a human voice. It is understandable, but it isn't totally clear. For example, my dog totally ignores everything Groucho says. Of course, this could be a function of my rather cheap speakers.

Sable

One problem with using text is that written text doesn't have a completely one-for-one relationship with the spoken text. In spoken language, there are intonations that are not represented in the text.

Sable is an XML markup language that was created to add the voice-specific additions to text. By using Sable, you can make the text pause, change the rate of speaking, and other speech modifications.

Perhaps even more important, you can specify the exact pronunciation of a word. This has uses in pronouncing things such as Scottish town names, saying last names, and choosing your preferred pronunciation for tomato.

You can also add sounds from sound files.

I haven't had the opportunity to use Sable yet, but I'm sure I will in a future robot.

Alternate sound output

So far, I've talked about text to voice. Although this is good enough for communication, you sometimes need more. Luckily, Linux has players for several types of audio files. If you have some theme music on MP3 or an AVI file of R2D2's beeps, your robot can play them. I would use the *alsaplayer* application to play such things.

To be honest, I don't do this. I have no objection to it; I just have never had the time to experiment with this sort of thing. Nor do I want Groucho to roam the house playing loud music. Ssomeday I might. I'm keeping my options open.

Speech to Text

For speech to text, I use Sphinx, which was developed at CMU. I was using Sphinx2 (which is available as a Gentoo package), but I recently changed to Sphinx4 (Figure 6-1). Sphinx4 is nice for me because it runs on multiple platforms easily. This means that I can test my code wherever I happen to be. Some people say that Java isn't fast enough to deal with speech, but they're wrong. Sphinx4 seems to perform as well as Sphinx2 (if not better) and is easier to configure.

Simple grammars (JSGF)

I admit that I cheat. I created a simple grammar of phrases that Groucho can understand, and I use the easiest way I've found to get it into a form that Sphinx4 can understand. I use the Java Speech Grammar Format (JSGF), which is somewhat

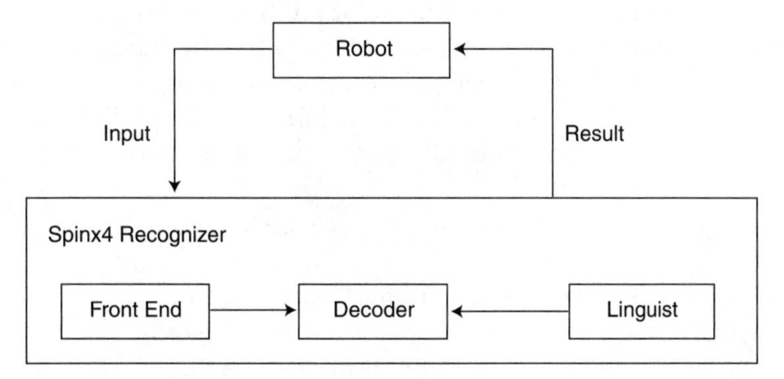

Figure 6-1 The Sphinx4 recognition system.

like BNF grammars. This is much easier than doing it the hard way.

The code I use will send only the text of strings that match the grammar.

The grammar I am using currently is defined as follows:

```
#JSGF V1.0;
/**
 * JSGF Grammar for Groucho
 */
grammar groucho;

public <fullCommand> = <botName> <command>;
<botName> = groucho;
<command> = <turnCmd> | <speedCmd> | <gracieCmd> | <ziantháCmd>;
<turnCmd> = turn (right | left);
<speedCmd> = (forward | reverse) [ <speed> ];
<speed> = slow | full | (<numberTo100> [ percent ]);
<numberTo100> = <digit> | (<digit> <digit>);
<digit> = zero | one | two | three | four | five | six | seven | eight | nine ;
<gracieCmd> = say goodnight groucho ;
<zianthaCmd> = say hi to zee ;
```

Some strings that are defined in this grammar are:

- Groucho turn right.
- Groucho forward 100 percent.
- Groucho, say "Good night, Groucho."
- Groucho, say hi to Zee.

The basic definition of JSGF files is fairly simple, and the full documentation is available at http://java.sun.com/products/java-media/speech/forDevelopers/JSGF.

An explanation of Groucho's grammar follows.

The grammar file must start out with the line:

```
#JSGF V1.0;
```

The grammar is named with the line:

```
grammar groucho;
```

The main rule is defined with the line:

```
public <fullCommand> = <botName> <command>;
```

Names surrounded by angle-brackets (“<...>”) are rule names. The line above means that <fullCommand> is matched by any utterance that matches the rule <botName> followed by any utterance that matches the rule <command>.

Terminal tokens are actual utterances. The rule for <botName> is defined as:

```
<botName> = groucho;
```

This means that the rule <botName> only matches the sound of the word “groucho.”

Rules can be defined as a combination of other rules or terminal tokens:

```
<speedCmd> = (forward | reverse) [ <speed> ];
```

The line above means that <speedCmd> is defined as the utterance either “forward” or “reverse” followed optionally by anything that matches the rule <speed>.

This seems very easy to me because the grouping is very much like the one used for regular expressions or BNF grammars.

Doing it the hard way: Training Sphinx

I admit that I haven’t done this. I’ve never had the time to try it yet. The main thing is that you use your own voice to train the system. This requires test utterances similar to what the robot can expect to hear. You need 30 minutes or more. Then you have to go through the audio files and mark the beginnings and ends of words. After this, you run the resulting files through a number of programs in order to get the proper probability of different sounds being in relationships to one another.

To do this well, you need to have many examples of different voices saying the same test sentences. This is how you make a good language definition file. Many of the current systems use files made by universities, which have many different test subjects.

The advantage of using a large group of people is that the voice recognizer will be more general and understand new people better. The advantage of using only your own voice is that your robot may recognize only your voice. You pays your money and takes your chances.

From Sphinx to Groucho

Converting speech to text is wonderful. However, I had to program a way to allow voice input to fit into my robotics framework. I chose to create a Sensor class to deal with this. I call it the StringSensor:

```
public class StringSensor extends AbstractSensor {
   public StringSensor(SensorImpl newSensorImpl, int newReadInterval) {
            super(newSensorImpl, 0, newReadInterval);
   }

   public void setValue(int newValue) {
            sensorValue = newValue;
   }

   public void setObjectValue(Object o)  {
            if (o != objectValue) {
                     objectValue = o;
                     setChanged(true);
            }
   }

   public String getString() {
            String s = (String)objectValue;
            objectValue = null;
            return s;
   }
}
```

This sensor expects a SensorImpl that can get a string from somewhere. I have a SensorImpl implementation that does this. It is used for Sphinx and opens a socket and returns all nonempty strings. This implementation is fairly simple, and strings can be missed if the Sensor is not called often enough:

```
public class SocketSensorImpl extends AbstractSensorImpl {
   protected String host = null;
   protected int port = 0;
   protected Socket socket = null;
   protected BufferedReader reader = null;

   public SocketSensorImpl(String h, int p) {
           this.host = h;
           this.port = p;

           openSocket();
   }
   /**
    * openSocket
    * <p>
    * Open the socket to the server if it isn't already open
    */
   public void openSocket() {
           if ((socket != null) && socket.isConnected()) {
                   return;
           }

           System.out.println("Opening socket");

           // Now open the socket
           try {
                   socket = new Socket(host, port);
                   reader = new BufferedReader(new
InputStreamReader(socket
                                                 .getInputStream()));

           } catch (UnknownHostException e) {
                   e.printStackTrace();
           } catch (IOException e) {
                   e.printStackTrace();
           }
   }

   public void readSensor() {
           String s = null;
           openSocket();
           try {
```

```
                        if (reader.ready()) {
                                s = reader.readLine();
                        }
                        if (s != null) {
                                s.trim();
                        }
                        objectValue = s;
                        intValue = ((s == null) || (s.length() == 0)) ? 0 : 1;
                } catch (IOException e) {
                        e.printStackTrace();
                }
        }
}
```

Sphinx wrap-up

I have found that I can speak within 10 feet of Groucho and be heard. The only disadvantage is that I have to speak the command continuously, without a pause after the initial *Groucho*. I would like to find a way to include the pause so that it sounds more like I'm getting Groucho's attention.

The microphone I use works fine. The system might work better with another microphone, but I haven't had the chance to check this yet.

Summary

- There are open-source programs that can handle speech.
- Festival is used for text to speech.
- Sphinx is used for speech to text.

Chapter

7

Vision: Seeing the World

Although a camera is a type of sensor, it has enough major differences from other sensors that it merits its own chapter (actually, it merits its own book):

- Vision consists of multiple images over time.
- Vision can be used at greater distances than a simpler sensor can.
- Vision allows a robot to "see" things the way a human would.

Each of these factors presents its own special problems and advantages. One of the major advantages of using vision is that it makes the robot more human-friendly. However, the biggest advantage of vision is that it allows greater distance of sensing and better generalized object recognition.

However, this chapter has space only for the very basics of image processing. This is something that people have devoted entire books to, and this is only a chapter. Here, I use vision for object avoidance rather than object recognition. In the future I plan on using vision for mapping, area recognition, and object recognition, as well as other things.

Luckily for me (and you), there are a lot of open source or available image-processing libraries out there. I am using a great number of them in Groucho for vision processing. Although I enjoy coding as much as the next programmer, I prefer to use other people's coding when possible.

Theory

Unlike simple sensors, vision requires a lot of preprocessing to make some sense of the images. We think of it as easy because we do it so easily. It seems that vision is a perfect application of artificial neural networks (ANNs). However, the large number of pixels that must be processed makes an ANN slow and difficult to train.

However, many objects can be recognized from their outlines, and most human-made objects have strong horizontal and vertical lines. Many artificial objects can be recognized by the relationships between those lines. Those relationships can be the same from a variety of angles. If one stores the relationships from only a few different angles, it is possible to recognize the object. This theory was the basis of the book *Knowledge-Based Vision-Guided Robots* by Nick Barnes and Zhi-Qiang Liu.

It is possible to differentiate objects by the relative lengths and positions of the line segments that compose them (Figures 7-1 and 7-2).

Three basic operations are used to extract outlines.

Figure 7-1 A chair before processing.

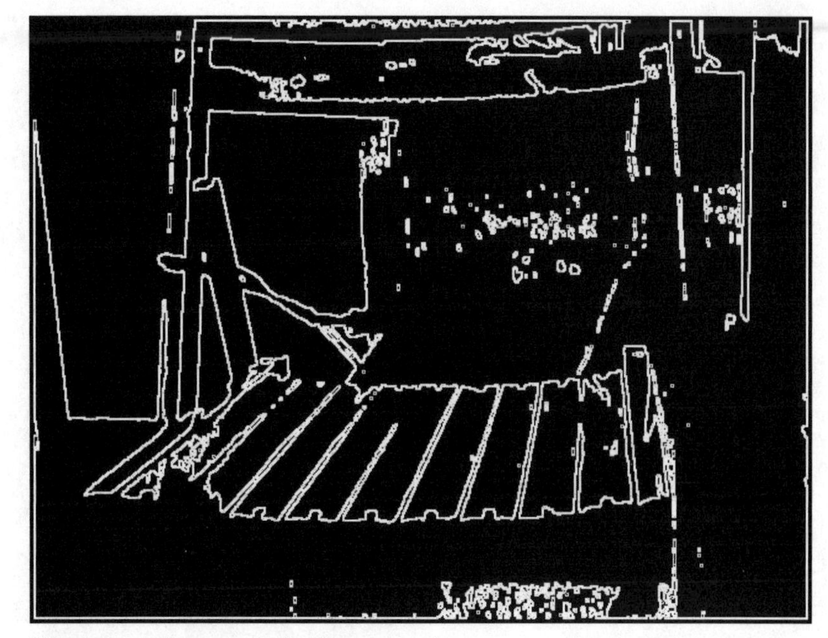

Figure 7-2 The same chair after processing.

Color mapping

By changing the color map of an image, you can make it much easier for more complex operations to work. Before I do anything else to an image, I do three color operations:

- **Grayscale conversion.** This operation converts a color image to a grayscale image by using the intensities of the individual pixels (Figure 7-3).
- **Histogram equalization.** This operation evens out the values of the pixels so that the image has a more level contrast (Figure 7-4).
- **Blobbing.** This operation removes all pixels except those that are close enough to a target color. This is used to aid in the recognition of objects of a known color.

Thresholding

With thresholding, you reduce the number of colors in the image. This is a simple operation and can be done without making a copy of an image.

Figure 7-3 A chair in grayscale.

Figure 7-4 A chair with histogram equalization.

Figure 7-5 The chair with thresholding.

Basically, you go through the image pixel by pixel and, if the value of a pixel is greater than the threshold, you increase that pixel to the maximum value; otherwise, you reduce the pixel to its minimum value. This can be done on the whole pixel or on each component of a color picture separately (Figure 7-5).

Variations on thresholding involve multiple thresholds.

Convolution

The third major operation is convolution. This is a multipixel operation in which a pixel's value is determined by the values of the pixels around it. Convolution requires making a working copy of an image to modify. Most convolutions are done with a square matrix of pixels in which the middle pixel in the square is the one that will be modified.

The kernels (matrices that are used for the convolution) normally will be three-by-three squares. However, it is possible to use any odd size for each dimension. As you can imagine, the computation goes up very fast for larger kernels. A 3×3 kernel causes 9 operations for each pixel; a 5×5 kernel causes 25 operations.

The rules behind a convolution are simple:

- Place the kernel so that the center is on the pixel to be modified (every pixel will be modified in turn).
- Multiply each pixel by the corresponding kernel value.
- The center pixel is equal to the sum of the new values.

A simple kernel to average pixels with the surrounding ones and blur an image is

$$\begin{bmatrix} 1/9 & 1/9 & 1/9 \\ 1/9 & 1/9 & 1/9 \\ 1/9 & 1/9 & 1/9 \end{bmatrix}$$

This simply replaces the center pixel with the average of the center pixel and all the surrounding pixels. This blurs the image by averaging each pixel with the surrounding pixels.

Another blurring kernel is to use a Gaussian function to determine the values for the array. The Gaussian function also is called the normal probability distribution. The smallest useful Gaussian kernel is 5×5.

$$\begin{bmatrix} 0.02 & 0.08 & 0.14 & 0.08 & 0.02 \\ 0.08 & 0.37 & 0.61 & 0.37 & 0.08 \\ 0.14 & 0.61 & 1.0 & 0.61 & 0.14 \\ 0.08 & 0.37 & 0.61 & 0.37 & 0.08 \\ 0.02 & 0.08 & 0.14 & 0.08 & 0.02 \end{bmatrix}$$

One thing I haven't talked about is what happens when the center pixel is on one of the edges of the pictures. Different situations require different answers. You can assume that the unknown pixels are all black or all white. Another solution is to assume that the unknown pixels have a value equal to that of the center pixel. I don't really care because my libraries will take care of this; it doesn't keep me awake at night.

By the way, my favorite method is to use the value of the center pixel to make up the missing data.

Hardware

First, I will discuss the imaging hardware that is available to me.

Groucho currently has a Logitech Orbit webcam. I bought this camera because it looked cool. I liked the idea of a camera on a stalk that would allow Groucho to "see" from a higher viewpoint. I could picture the camera coming up and panning to the left and right (with the appropriate "zipping" noises). Unfortunately, this didn't work out because when I mounted the camera on its extension, the camera wobbled too much.

Once I removed the extension and found a reliable driver, the camera worked fine. However, the Orbit is a USB 1.0 device rather than USB 2.0. As I'm writing this book I still can't find a good USB 2.0 camera and driver combination for Linux.

I was going to use an AVRCam, which is a simple camera module that is processed by an AVR microcontroller and communicates to the motherboard via basic RS232 commands. One nice thing about this camera is that it has preprogrammed routines that allow you to get quick information, such as the position of color "blobs" so that the camera can be used to follow objects of known colors. The computer doesn't have to know the image so the data transfer is much faster. The main disadvantage of the AVRCam is that it is only available as a kit.

A similar product is the CMUCam. It was developed at CMU and is sold at many robot stores on the Web.

I prefer the AVRCam because the programming is open source. It is possible to add to or change the original programming of the camera. This means that it may be possible to do some of the image processing at the camera level.

I do have the AVRCam mounted on Groucho, but I couldn't get it to do what I wanted in time before this book was finished. Thus, my vision experience is with the Logitech Orbit webcam. In the future, I will use the AVRCam to locate color blobs and the webcam to determine more detail.

Some software in my robotics framework deals with the AVRCam, but I'm not sure that it works at this time.

Software

You need both low-level and high-level software to get images from the webcam to the application.

Driver software

Low-level software for getting the images with my camera requires the following:

- The kernel must be compiled with the Video for Linux (V4L) options on.
- The PWC (Phillips WebCam) driver. This is available as a Gentoo package (*usb-pwc-re*). This is a completely open-source fork of the original PWC driver; the original driver was pulled from the Linux kernel because of differing opinions. The new driver is also available at http://www.saillard.org/linux/pwc.
- The Camsource V4L image server. This is also available as a Gentoo package (*camsource*). This is a frame grabber and image server for V4L drivers. The source is also available at http://camsource.sourceforge.net.

High-level software

Luckily for us, there are a number of different open-source libraries that deal with high-level vision processing. Some of them are easier to work with under Linux than others:

- **MAVIS.** This is a vision library written specifically for robotics as part of the LEAF project. Currently, it works under Windows only, but as soon as I finish writing this book, I will be helping to port this to Linux. Currently, it is available at http://www.leafproject.org. To be honest, MAVIS is beyond anything I've accomplished so far.
- **Gandalf.** This is an image-processing library that works under most UNIXx-like platforms, including Linux. It has all the basic image-processing elements you'd ever want. It is available at http://gandalf-library.sourceforge.net.
- **Java 2D.** This is a core part of Java, which has most of the operations that are needed to do image processing. It is part of the base Java system.
- **Generation5.** This is a Java library that was designed for robotics, and it includes an extensive vision package. It is available at http://www.generation5.org.

I created a simple abstraction class (ImageMinder) so that I can use whatever library I want to use at any time. Currently, I'm using Generation5, but I will switch to MAVIS when I get the time.

Feature Extraction

The idea of feature extraction is to bring out the shapes and features that are important. These are things that our brains "instantly" point out to us. Looking around my living room, I see a TV, an entertainment center, a birdcage, and a couple of hospital beds.

I don't know how my brain does this so quickly. I have read a lot about this, but there are many competing ideas, papers, and theories. Part of the problem is that human vision is a complex process that goes from the eyes to the brain through several layers of extremely complex neural networks.

In addition to finding objects, I can identify familiar objects so quickly that I don't even notice the process. I can't do this on a computer yet. However, the problems are mostly in the software.

Edge detection

We identify objects by various properties. Perhaps the most important of these properties is shape, and a shape is defined by its edges.

One of the main steps along the way is to locate the edges in each of the images. This isn't exactly a trivial process. To keep the processing time down, I'm using an image size of 80 by 60 pixels. If you think about it, a standard VGA image is 640 by 480 pixels, which is a total of 307,200 pixels. The smaller image has only 4,800 pixels, or about 1.5 percent the number of pixels. This makes the process of image processing *much* easier.

The steps needed to locate the edges in this picture are as follows:

- Start with an original image.
- Convert the image to a grayscale image.
- Apply histogram equalization to the image. This changes the color map of the image so that the pixel values are spread more evenly between white and black.
- Apply a Gaussian filter to the image. This is a 5-by-5 convolution matrix that is based on a two-dimensional Gaussian function.

- Apply a two-point thresholding operation to convert the image to a black-and-white image (with nothing in between).
- Apply the Sobel edge-detection algorithm.

The resulting image is, one hopes, black everywhere except for the edges.

Histograms

Since the picture is now composed of a black background and white pixels for the detected edges, it is possible to make histograms for each of the horizontal and vertical lines. By checking the values, it is possible to tell if a particular line is mostly horizontal or vertical. This process isn't perfect, but it can be used to show strong vertical edges (Figure 7-6).

This can be a quick and dirty way of seeing some classes of objects. Most artificial objects have strong horizontal and vertical components. For a robot, I think that vertical lines are more important because they indicate immediate obstacles.

By counting the number of white pixels in each column, you can get an estimate of the number of vertical edges in that column.

Another way to use histograms is to use each of the histogram values as input into an ANN. With a picture size of 80

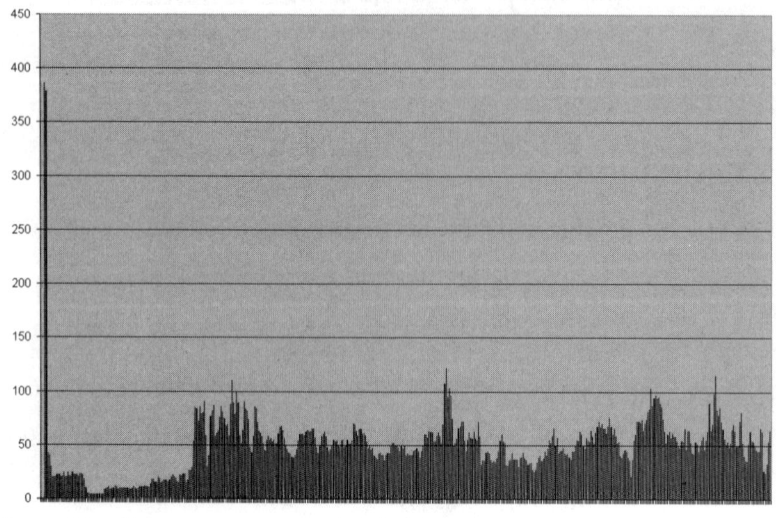

Figure 7-6 Vertical histograms of the chair.

by 60 pixels, this gives a total of 140 inputs (assuming both horizontal and vertical histograms) to the network. I would suggest an output of 16 probabilities that divide the camera's field of view evenly. These outputs will be higher if there is an obstacle corresponding to each output.

Optical Flow

Many of the ways people navigate in the world involve vision. When driving, for example, we estimate our speed and rate of turning by seeing the apparent motion of the background. This is very similar to the way optical mouses work.

Optical flow (Figure 7-7) is extremely important to the way people navigate. If the flow is roughly the same out of both sides of our peripheral vision, we are gong straight. If we turn, the flow increases on the outside side of the turn.

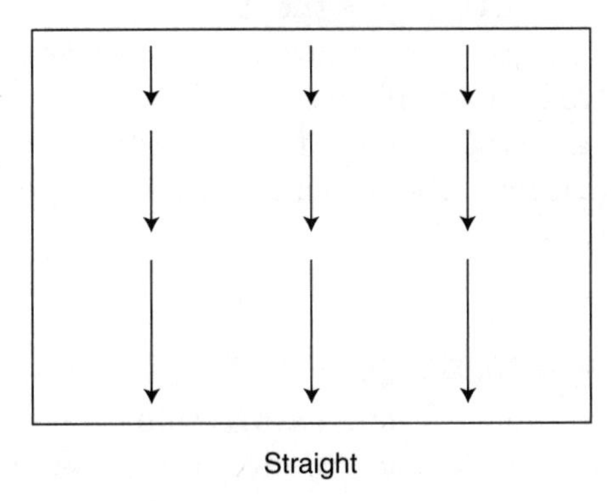

Straight

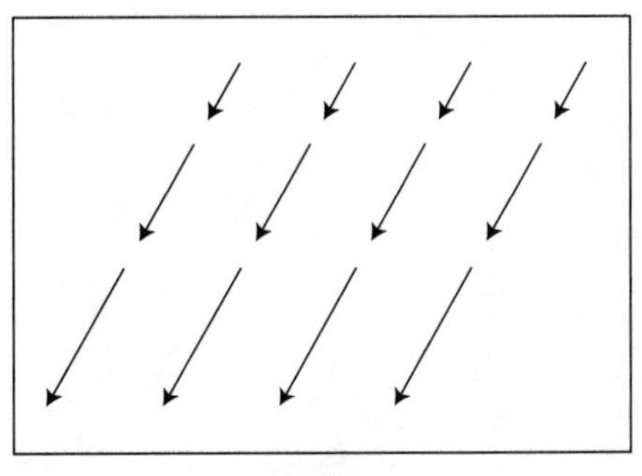

Turning Right

Figure 7-7 The optical flow changes when we turn.

Possible algorithms

One way to estimate the optical flow is to use a few rows of the image (after Gaussian filtering), say, five rows one pixel in height spread evenly from the top to the bottom. From these rows, take five sets of five pixels spread evenly from the right to the left (Figure 7-8).

Another possibility is to take a number of single pixels at set positions scattered near the center of the image (maybe a total of 25 pixels) and use them as input to an artificial neural network (ANN). The output could indicate the speed and rate of turning. This would require a network with feedback, so that the changes in the value over time will be represented (Figure 7-9).

Normally, it would require a lot of time to train this network manually. However, it is possible to use the information from the odometers and the sonar units to train the network automatically. For example, I can use the odometers to get the current speed and rate of turn, and train the network at this time. After a training period, I can use the sensory information to double-check on the accuracy of the ANN.

Optical flow can give you information, but sometimes the view shows contradictory information. This can happen when an object moves across the field of vision. When an object contradicts the optical flow, the object is moving (Figure 7-10).

Color Blobs

A simple way to produce vision is to look for objects of a specific color. For example, if your living room walls are painted pink, just seeing a pink background may be enough for the

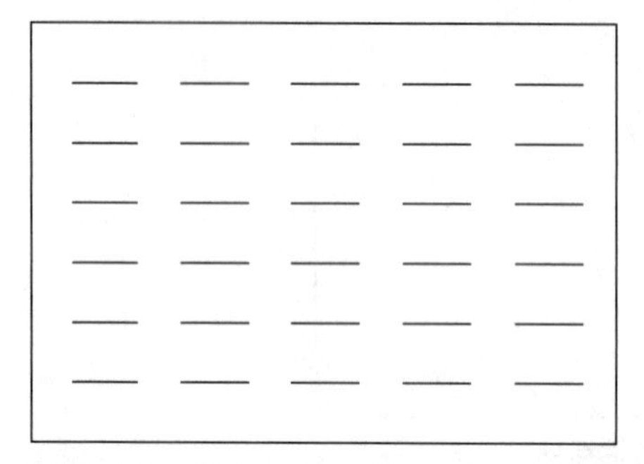

Figure 7-8 If one measures the changes of the brightness of the specified sets of pixels, the optical flow can be estimated quickly.

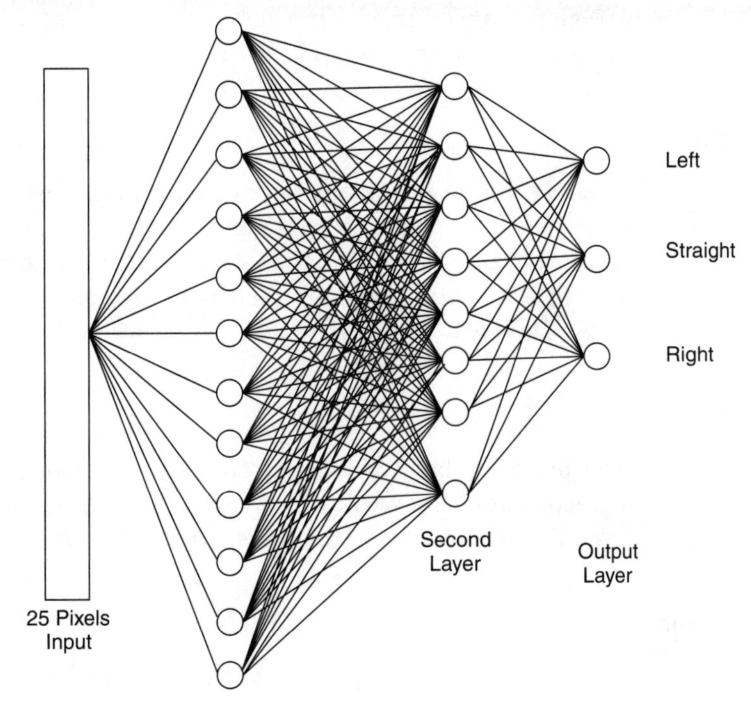

Figure 7-9 One possible ANN configuration.

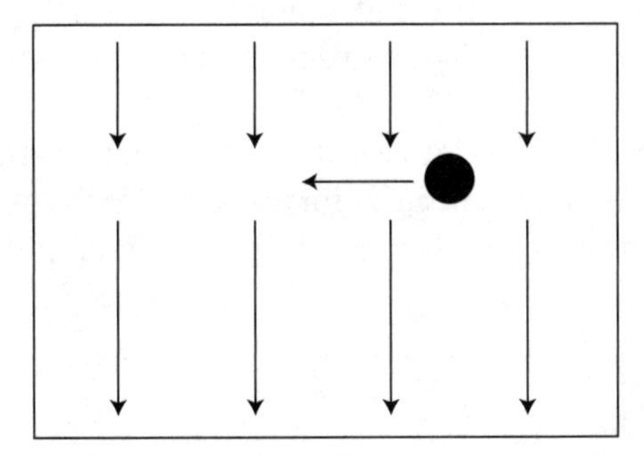

Figure 7-10 Object moving against the optical flow.

robot to know it is in the living room. Basically, you go through each pixel and keep the pixels that are close enough to the target color.

Signs and Beacons

So far, there has been no modification to the environment. However, sometimes, just as road signs are useful to people, some signs can be useful to robots. There are several ways to make signs that are readable by a robot.

Plain text

As long as a sign can be made obvious enough, perhaps by creating the sign with specific background and foreground colors, optical character recognition (OCR) algorithms can be used. There are several open-source OCR programs available, such as *OpenOCR* and *Waygoer OCR*.

Bar codes

If you print the bar codes large enough, they can be read with the resolution of a webcam. Again, there are several open-source libraries to read bar codes, such as *readbarcode*.

Bicycle reflectors

If you have only a few places that have to be located, you can use different colored bicycle reflectors for your beacons. One easy way to locate doorways is to use two colors of reflectors, perhaps red and green, and put one on each side of each doorway consistently. I would put green on the right side and red on the left side (Figure 7-11).

For additional credit, you can use a number of different colored reflectors for additional information. Perhaps you could use yellow, orange, and blue reflectors. If you placed two of

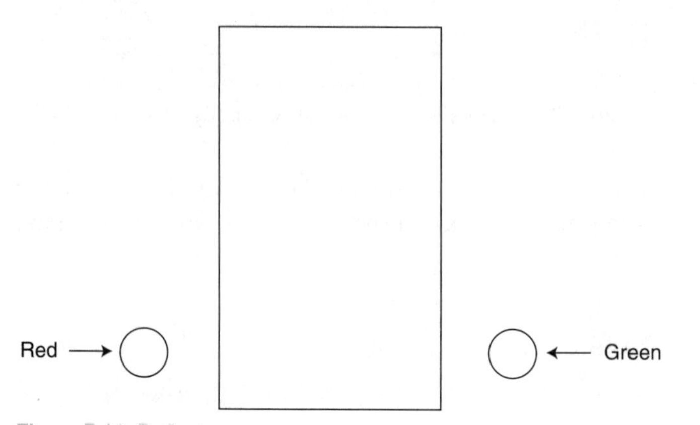

Figure 7-11 Reflectors.

them in a specific relationship to the other reflectors (perhaps placed 2 inches above the green reflector), they could add more information. By using two reflectors chosen from the three suggested colors, you could identify nine different areas.

Existing signs

Since many people, including myself, are not able to add robot-friendly signs to our houses because of spousal veto, we have to make do with what is there. Some existing signs that may exist in a house are:

- **Different colored/textured floors.** My kitchen, my bathroom, and a small square near the entranceway have linoleum; the rest of the downstairs has brown carpeting.
- **Different colored walls.** Although my house may have all off-white (I can't remember which of the many shades of off-white the walls are), some houses have more creative painting schemes. I really need a bigger house so that I can be more creative in wall colors.
- **Distinctive furniture.** My dining room has large bookshelves and floor-to-ceiling curtains; my living room has a brown birdcage, the entertainment center, a couple of beds, and a dog bed.

Summary

- Vision consists of examining a large number of pixels quickly.
- Image processing is a major part of robotic vision.
- Vision can be used in place of other sensors.

Chapter

8

Mapping and Planning: Where Is Here? and How Do We Get from Here to There?

I warn you at the start: Some of this chapter concerns things related to symbolic AI. This shouldn't scare you off. It also shouldn't be a surprise, because a map is a symbolic representation of the world. Trust me: I won't go into theories and Greek letters. If you want those, there are a number of large books and theses on the subject. Feel free to send my editor nasty e-mails if you find a single proof in this chapter.

For simple maneuvering, a map is not really necessary. One of the first types of nontrivial bots that we build is what I call wallbangers. These robots use a very simple behavioral algorithm. That algorithm can make a robot appear quite intelligent, but in reality this is trivial.

Another common behavior that doesn't require a map is wall following.

I mention these behaviors specifically because they can aid you in making a map.

There are two questions a map can answer:

- **Localization:** Where am I now?
- **Navigation:** How do I get from there from here?

I should say now that there are many other ways to navigate without a map. Most of these ways involve placing some sort of marker in the environment. Some examples are

- Reflectors (yes, basic bicycle reflectors) of varying colors
- Active beacons, typically using some sort of pulsed IR
- Optical or magnetic strips in the floor

Unfortunately, all these techniques require changing the environment to work. Sometimes this is difficult or even impossible, and sometimes the environment changes too quickly to allow the placement of markers.

However, in many cases, we want to build and use a map of the world. This can be for many reasons:

- We want the robot to choose dynamically the routes to the destination (route planning).
- We want to have the robot act differently in different areas.
- We want to create a human-readable map of an area, building, and so on.
- We want to be able to tell the robot to go to a specific place.
- We want the robot *not* to go into certain areas (for example, it would be very bad and somewhat expensive if Groucho tried to go down the basement stairs).

For my purposes, I rarely care if the map is completely readable by humans. I just want to say, "Groucho, go to the dining room" or "Groucho, find the Bug." I will admit that sending a robot to find a dog is rather pointless because dogs can climb stairs and my robots can't (I'm thinking about it) and, so far, Bug is extremely scared of all my robots. Therefore, the map and I must agree on a few landmarks, but I really don't mind how the robot represents things.

Simple commands such as "Go north" are fairly easy. However, "Go to the dining room" is more difficult. The following questions have to be asked in order:

1. Am I already in the dining room?
2. If not, where am I?
3. How can I get to the dining room from here?

In my framework, I have objects that deal with this:

1. The *mapper* does the work of creating and updating a map.
2. The *locator* answers the question Where is here?
3. The *navigator* answers the questions Where is there? and How do I get from here to there?

I keep these objects separate because there are many ways to keep a map and many ways to compute the path between two points.

Mapping Algorithms

Several algorithms are known to be useful in mapping. There are many that I don't know. I will concentrate on the easier ones (the ones I know). You can use almost any combination of sensors to map an area. One robot, the Cye, uses only feedback from the wheels for mapping. Groucho uses a combination of odometry and sonar readings for mapping.

Topological maps

No, this isn't a typo. A topological map (Figure 8-1) keeps the relationship between places without specifying a strong metric (that is, distance isn't measured precisely or even at all). In other words, it might say that the living room is next to the dining room without specifying how long the distance is. This gives the information for getting from one place to another without specifying a complete map.

The human brain seems to contain many maps that work this way. Anybody who has heard the cry "Are we there yet?" from the back seat can attest to that. Most people seem to be better at knowing relationships than at understanding quantities. This seems to work fairly well for us. We can use the relationship data of a topological map to get close to our destination, and use our senses to deal with the actual distances involved.

In computer programming, a topological map might be represented by a collection of nodes, where each node is a place. There will be paths between nodes that have some of the information used to travel between the nodes. As the robot travels the area more completely, the information for the path can be

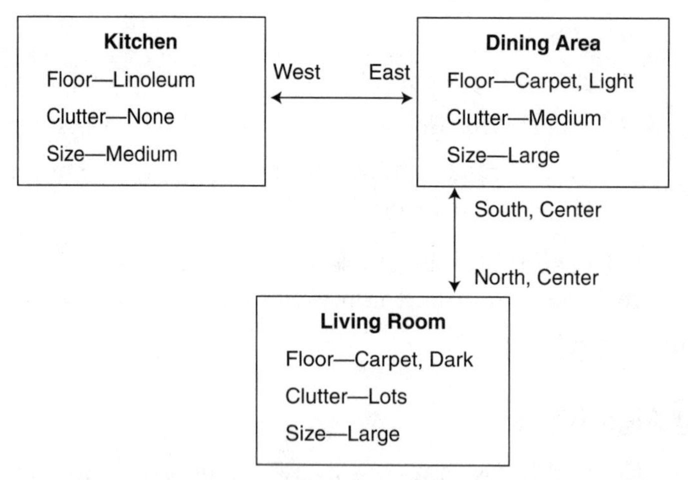

Figure 8-1 A topological map.

made more detailed. As new areas are explored, new nodes and paths can be added. If an interesting place is found while one is traveling along a path, a new node can be created, and the path can be broken in two.

Occupancy maps

An occupancy map (Figure 8-2) is closer to a written map. A grid is created in the robot's internal memory, and then individual grids are marked with different states: occupied, free, and unknown. There are many variants of the basic grid map.

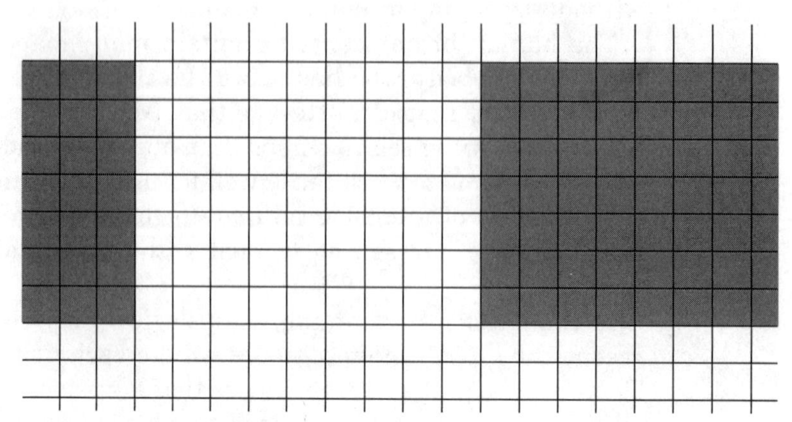

Figure 8-2 The black squares are occupied, and the white squares are unoccupied.

Grid maps are nice for programmers because a grid is easily mapped by a two-dimensional array.

A grid doesn't have to contain merely the three options above. If you want to be more detailed, you can have an object associated with the obstacle that will give more information about it.

One simple variant of the grid map is to have a map in which the grid cells are fairly large, and the robot can subdivide any cell into a smaller grid map (Figure 8-3). This way, large, empty, and unexplored space can be supported easily, and complex areas can be mapped in more detail. Of course, my house is small enough that this is unnecessary.

Even more important, a grid cell can contain information about the area covered by the grid cell:

- The texture of the floor
  - Carpeting
  - Linoleum

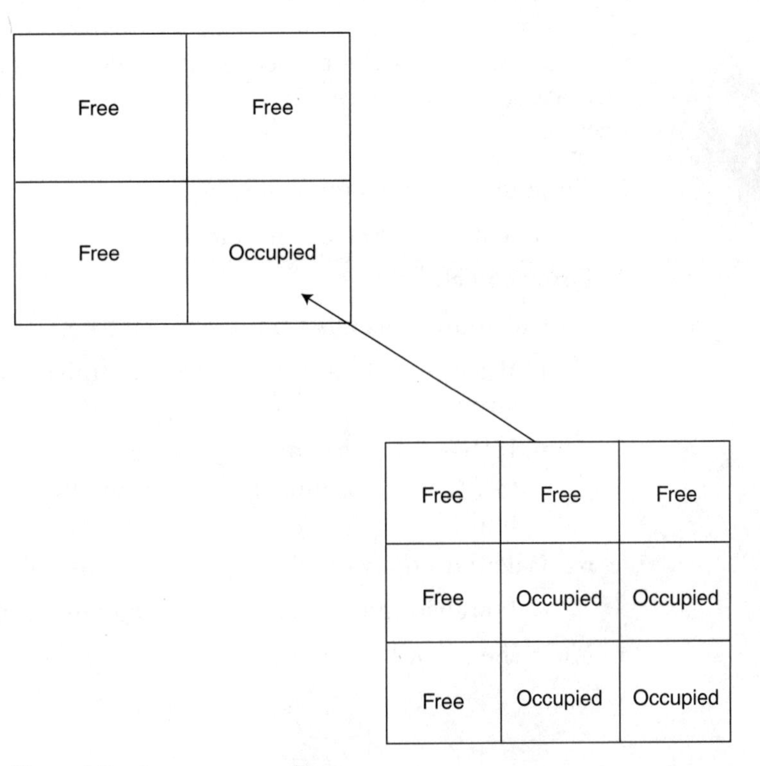

Figure 8-3 An occupancy submap.

  - A hole
- The chance of a movable obstacle being there
  - A person
  - A dog
  - A kitchen chair
- The type of obstacle
  - Height
  - Appearance to sensors
  - Connection to obstacles in adjacent cells
- Areas where the robot is not allowed to go

For mapping my house, I chose both a starting and a minimum grid of 10 cm square. This size is easily measurable by Groucho, and I can fit the complete map into memory easily. I could have chosen a starting grid size of 100 cm and created submaps when necessary. However, I live in a small house and know that my map will fit easily into memory. If I were to map a larger area with big empty areas, I would use a larger starting grid size so that the empty areas would take less memory.

Creating a basic occupancy map is a fairly easy iterative process.

1. Get a quick polar map from the sonar units.
2. Assume a 10-degree arc for each sonar unit.
3. For each echo:
   a. Calculate the cells that should be free.
      i. If these cells are marked as occupied, something is wrong.
      ii. If these cells are already free or have a low occupancy probability, multiply the occupancy value by one-half.
   b. Calculate the cell where the echo comes from.
      i. Increase the occupancy (to a maximum of 100).
4. Move the robot by about 10 to 20 cm and repeat step 1.

Polar maps

A polar map is a map represented by a circle, with the robot at the center. Obstacles can be represented by angles and distances from the center.

Grid maps are easy to create and understand, but if you're using a map from the point of view of a robot, a map based on polar coordinates may be more useful. Sensor readings normally are taken in an arc or circle about the robot; therefore, the data already are in this form. To determine the optimal direction to head, the computations are sometimes easier on a polar map.

A polar map can be constructed very quickly from sensor data and used to determine information about the immediate area quickly. The sonar units on Groucho can construct easily such a map (Figure 8-4).

There are a couple of uses for such a map. First, it can be used to determine quickly which directions are free of obstacles. Second, as stated previously, this information can be used to create an occupancy map. Third, it can be used to locate the robot within an existing occupancy map. By finding areas on the map that fit with the polar map, it is possible to estimate the robot's location. If there is no preexisting location information, the number of potential matches may be large, and the robot will have to move to a different area and take further

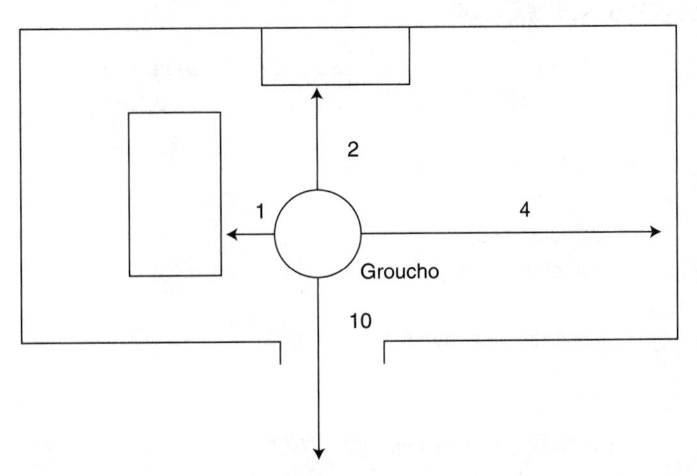

Figure 8-4 For simplicity's sake, I show only four directions from the sonar data. Groucho can easily navigate while using only the sonar data.

measurements. If the robot was at a previously known location, the number of potential matches should be small.

It is even possible to use a polar map of the immediate area to help with robot localization—that is, determining where the robot is on the main map. However, this probably will involve the robot moving and taking readings in a few places to determine the position absolutely.

Well-known locations

Every now and then a location is found that is fairly easy to find again. This is termed a well-known location. When your robot is in such a location, it can be reasonably sure that the location on the map is known.

It is possible to navigate from one well-known location to another. This makes it easier to deal with the real world.

Such locations can be locations that are obvious in the real world, such as doorways and right in front of the stairway. These locations also can be visible only to sonar or infrared, or they even can be in visible light from the point of view of a 3-foot-tall robot.

Programming the Maps

It's easy to talk about maps, but it's another thing to get down to the code. Luckily, I've done some of this work already.

OccupancyMap and MapCell

The OccupancyMap class is just an array of MapCell objects. A MapCell object is a cell within an OccupancyMap object:

```
public class OccupancyMap {

    protected MapCell[][] map = null;
    protected Point2D cellSize = null;

    public OccupancyMap() {
    }

    public OccupancyMap(int x, int y) {
            this();
            map = new MapCell[x][y];
```

```
    }

    public void setMapAt(int x, int y, MapCell newCell) {
            map[x][y] = newCell;
    }

    public MapCell getCellAt(int x, int y) {
            if (map != null)  {
                    return map[x][y];
            }
            return null;
    }

    public Dimension getDimension() {
            Dimension d = null;
            if (map == null) {
                    d = new Dimension(0, 0);
            }
            else {
                    d = new Dimension(map.length, map[0].length);
            }

            return d;
    }

    public Point2D getCellSize() {
            return cellSize;
    }

    public void setCellSize(double x, double y) {
            cellSize = new Point2D.Double(x, y);
    }
}


public class MapCell {
    // Occupancy
    public final int UNKNOWN = 0;
    public final int OCCUPIED = 100;
    public final int FREE = -100;

    // Surface textures
```

```
    public final int SURFACE_UNKNOWN = 0;
    public final int SURFACE_MIXED = 1;
    public final int SURFACE_CARPET_SHORT = 10;
    public final int SURFACE_CARPET_LONG = 11;
    public final int SURFACE_HARD_SMOOTH = 20;

    protected OccupancyMap map = null;
    protected int surface = SURFACE_UNKNOWN;

    /**
     *
     */
    public MapCell() {
    }

    /**
     * getMap
     *
     * @return The OccupancyMap that may be in this cell
     */
    public OccupancyMap getMap() {
            return map;
    }

    public void setMap(OccupancyMap om)  {
            map = om;
    }

    public int getSurface() {
            return surface;
    }

    public void setSurface(int s) {
            surface = s;
    }
}
```

Mapper

The mapper object is merely a set of behaviors that implement a mapping algorithm. Basically, any set of behaviors that can move the robot in a predictable way can be used.

Locator

The locator class uses both the current OccupancyMap and sensory data to determine the robot's location. Assuming that there are no markers in the environment, you have to use both the sensory data and the last known location. Starting at the last known location, you can check the current sensory data against the map until you find a match.

If you don't have a previously known location, you first check to see if the sensory data indicate that the robot is in a well-known location.

In all cases, if there are multiple possible matches, the robot should move to another location that will differentiate the two locations.

Navigator

The navigator class tells the robot how to get from one place to another. Several very important variables are dependent on the robot. For example, the size of the robot and the size of the map's grid cells determine how close to an occupied cell the robot can safely get.

There are many possible algorithms to determine which route is chosen. This is why navigation is also called route planning:

- A flood-fill algorithm
- Going from one well-known location to another
- Using visible landmarks

Summary

- By keeping a map of the environment, a robot can navigate better in the environment.
- Occupancy maps show the chance that any given area is occupied.
- Polar maps are made quickly from sensory data.
- Well-known locations can be used as landmarks to aid in making the navigation self-correcting.

Chapter

9

Artificial Neural Networks

Artificial neural networks (ANNs) were developed originally as an attempt to understand biological neural networks. Then they were taken over by computer scientists, dropped, taken over by the biology guys again, and then picked up again by computer people (and so on and so forth).

Basically, there are two camps of people who use ANNs: those who are using ANNs to solve problems on a computer (the U.S. Postal Service uses a complicated ANN to handle the problem of reading addresses) and those who are using ANNs to understand how thinking occurs. Because I'm a generalist, I have a foot in both camps. Part of me wants to understand thought, and part of me just wants to create a robot that acts as if it thinks so that people can argue about whether it thinks or just imitates thought. And yes, I consider making large numbers of people think hard about a problem to be a good use of my time. Or, maybe I just find it amusing.

For this chapter, I'm concerned with the use of ANNs for computing. I'm not trying to simulate any biological processes.

There is also a camp of programmers who have called the ANN a magic bullet that can be thrown at a problem when one does not know how to program it normally. I agree with this viewpoint a little. There are a few problems that I don't know how to deal with properly, and so I throw an ANN at them. Groucho has the computing resources to deal with those problems.

The Artificial Neuron

At the base of any artificial neural network will be, of course, artificial neurons. Real biological neurons are extremely complicated. Those cells receive electrochemical input in the form of pulses from many (hundreds or thousands) of other neurons asynchronously. The output is in the form of pulses. All of this is determined by complex chemical reactions.

Since doing a detailed simulation of a large number of actual neurons would be extremely time-consuming (not to mention that it would require more detailed knowledge of the brain than we currently have), several simplifications are made. The main simplification has been to assume that we are integrating a number of pulses over a short period, and so we can convert the rate, duration, and strength of the pulses into a simple value.

The McCulloch-Pitts neuron

In 1943, the first artificial neuron was defined in a paper written by Warren McCulloch and Walter Pitts. Although this model is considered simplistic today, it is still used, and many modern artificial neurons are based on the McCulloch-Pitts neuron (Figure 9-1).

The basic rules that govern this neuron are as follows:

1. The neurons are calculated simultaneously.
2. The inputs are composed of either an original input or the outputs of other neurons.
3. Each neuron has a threshold value.

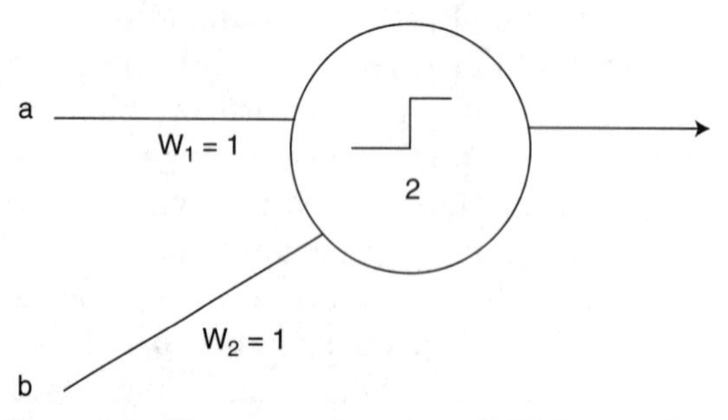

Figure 9-1 The neuron above is an "AND" gate: When both the inputs are 1, they add up to 2 and equal the threshold value so that the output equals 1.

4. The output is 1 if the sum of the inputs is equal to or greater than the threshold; it is zero otherwise

The generic neural network artificial neuron

Since 1943, the understanding of both neurophysiology and computing has come along. The McCulluch-Pitts (MP) neuron is too simple for many uses, but it is still employed today. However, with a few simple changes and modifications, a more powerful artificial neuron can be created (Figure 9-2).

1. Each input has a weight associated with it; the input value is multiplied by the weight to give the input value
2. The neuron also has a transfer function associated with it; this corresponds to the threshold value of the MP neuron.
3. The value of the neuron is the sum of the input values, with this total run through the transfer function

The exit function is traditionally a function that bounds the output while allowing the normal values (close to the mean) to pass through with little change. The sigmoid or the tanh functions work in this way. A basic threshold function also can work; however, certain training techniques require that the exit function be differentiable so that the more complex functions work better.

The sigmoid function (Figure 9-3) is defined as

$$\frac{1}{1 + e^{-x}}$$

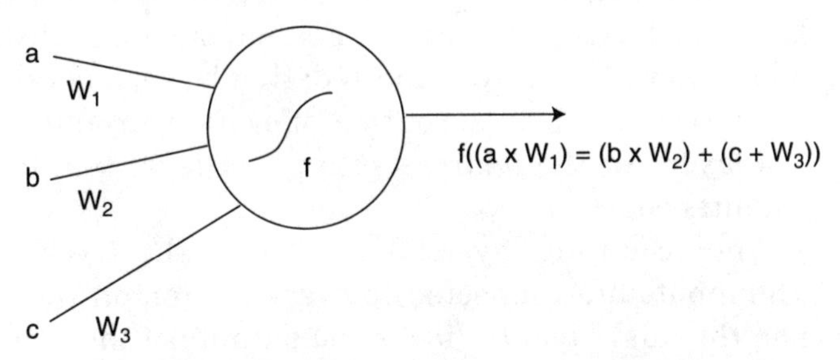

Figure 9-2 A more generic artificial neuron.

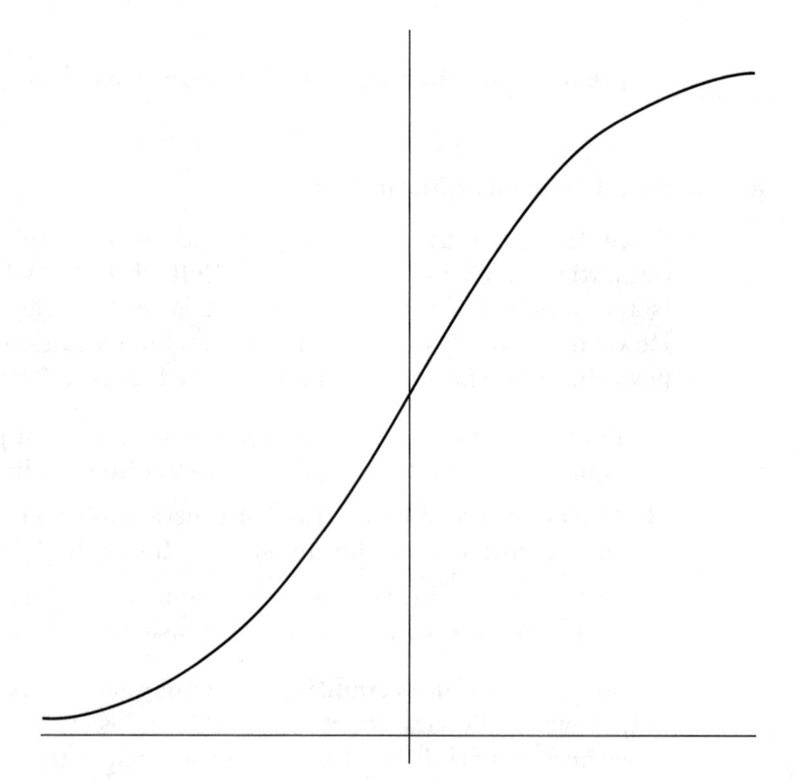

Figure 9-3 The sigmoid function.

Note that, if the sum of the input is very large, the output will still be limited to 1. If the sum of the input is very small, the output will be limited by zero.

Layers in an Artificial Neural Network

There are many ways to organize neural networks. Most of these methods are used to organize the neurons in layers. A layer is an array of neurons. For the purposes of this chapter, all layers will be fully connected; that is, each neuron in a layer will be connected to the output of every neuron from the previous layer. All the neurons in a given layer will be calculated simultaneously.

When counting layers, I count only the layers of neurons. The inputs are connected to every neuron on the input layer, and the final layer connects the outputs of the neurons directly to the output of the network.

This is a simplistic model of neural networks. It is possible to build networks in which specific neurons feed back into one input of a previous layer.

Single-layer artificial neural networks

For simple problems, sometimes simple solutions are the best.

A single-layer neural network (Figure 9-4) is composed of the layer of inputs, the layer of neurons, and the output layer. Each of the inputs is connected to each of the neurons. The output of each neuron goes directly to the output.

One reason single-layer networks aren't used much is that there are certain problems that a single-layer network cannot solve. For example, it is impossible to create a single-layer network that can handle the XOR function.

Multilayer artificial neural networks

To solve the problems associated with the single-layer ANNs, multilayer ANNs were created. A two-layer ANN (Figure 9-5) can approximate (to any required degree) any function, given enough neurons and connections. However, some problems are more easily and quickly solved with more than two layers. This makes an ANN a universal function approximator.

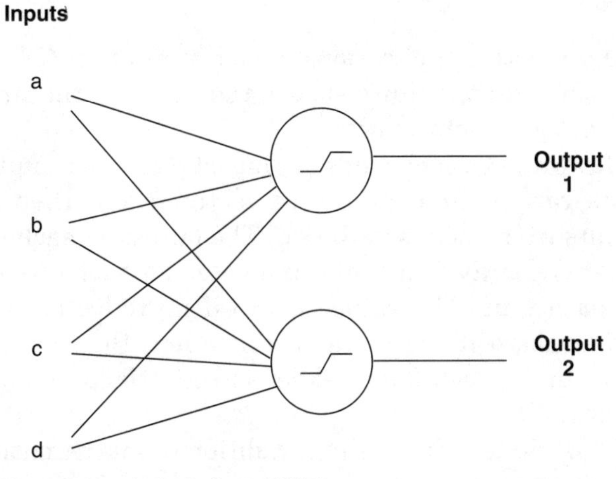

Figure 9-4 A single-layer ANN.

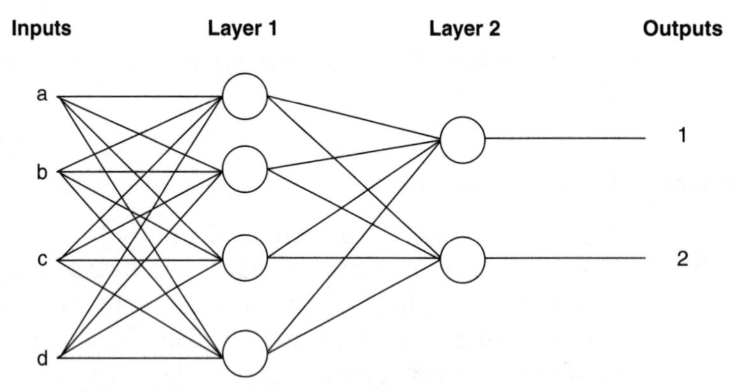

Figure 9-5 A two-layer ANN.

Neural Network Software

Several good neural network libraries run under Linux. Because my examples are in Java, I will be using Java Object-Oriented Neural Engine (Joone). One reason I like Joone is that in addition to the core ANN engine, it has a very nice interface that works under both Linux and Windows. Joone's help window contains much better instructions than I can give now, and so I won't go into that much detail.

Joone consists of two packages: the user interface and the core engine. Groucho needs only the core engine. My development machine needs the full kit. I develop on a Windows laptop and send the finished product to Groucho.

Sound and neurons

Many robotics applications do not require an ANN. For a simple sonar unit, I can measure the time for the first reflection and get good information.

However, Groucho has a ring of 12 sonar units. All those units can be fired at the same time, and then an array of results from each is gathered. The output of each sonar unit is a 32-byte array that contains a value for each 35 cm of distance for each unit. The value is zero if no reflection was detected and nonzero if there was a reflection. Because there are 12 sonar units, each unit will detect reflections from (perhaps) all 12 units.

To decipher this using traditional mathematics and programming is beyond me, but seems right for a neural network.

This gives a total of 384 inputs (12 sonar units times 32 bytes) in the input layer.

I decided to build a two-layer ANN with the second (hidden) layer having 32 neurons and the output layer having 8 neurons. The object was that each one of the eight neurons should be able to show the nearest obstacle in a given direction. Unfortunately, even after I constructed the objects used for this, I ran into a major problem: My house had become too cluttered and two small for this to work. The sonar units (SRF08s) have a resolution of only about 1 foot in artificial neural network mode, and my house doesn't have that many places with that much extra room. Basically, I get an array that is mostly 1s.

However, I will give you the steps I took to construct this layer in the hope that it will work in other places.

Joone

First, you need to download the full Joone package from http://www.joone.org. It is quick and easy to install.

You need a machine with a monitor, so that you can run the graphical editor. I run mine on my Windows laptop, it also can be run under Linux. Groucho has enough horsepower to run this, but I'm not running X on Groucho. Luckily, you can develop on one machine and run on another.

For your robot, you will download just the core Joone engine. This is the same as the previous download, without the bells and whistles (i.e., the graphic user interface).

Joone looks pretty much like a drawing program except that you are drawing ANNs rather than merely lines and boxes. Three toolbars with symbols are used for creating and editing ANNs. The help that comes with Joone is better than anything I could write.

The final neural network that I came up with is shown in Figure 9-6.

Summary

- ANNs are useful for some classes of problems.
- Most ANNs that are useful for robotics have multiple layers.

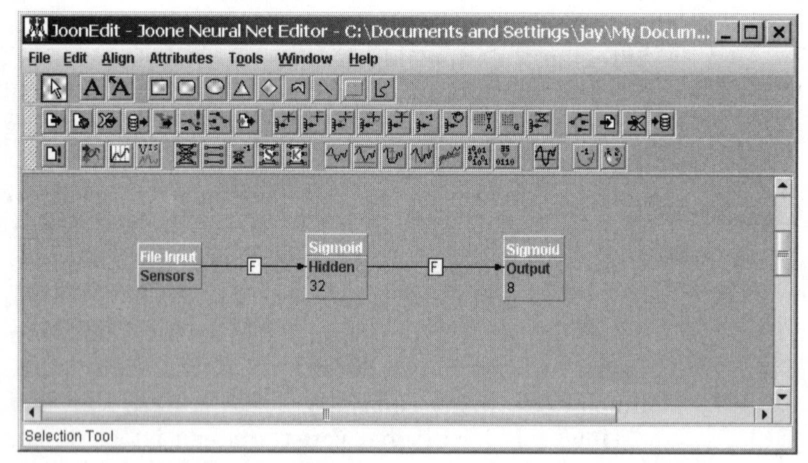

Figure 9-6 Joone.

Chapter 10

Working without a Wire

For a robot to be both mobile and autonomous, it must be self-contained. That is, there can be no wires tethering the robot to a fixed location. However, there is virtue in communication. Hence, this chapter discusses wireless communication.

To be honest, almost everything in this chapter, except for the hardware specifics, could apply to traditional Ethernet or other wired connections. The wireless part comes simply from the requirements of a mobile and autonomous robot. All the software I am mentioning in this chapter is open source and easily available.

There are several ways to do this, depending on the difficulty you are willing to endure, your budget, the range of communication, and the size of the robot. For Stuart, I took the easy route and used an Ethernet to 802.11b bridge. This is a simple device that has an Ethernet connection, a power input (5 volts DC), and an antenna—all of this with the circuitry to connect the parts. Once my robot is on, it becomes part of my home's WiFi network.

For Groucho, I found an easier solution: I have a Linksys 802.11g card. This card uses the rt2500 driver, and this is not a part of the Linux kernel yet. However, it is a Gentoo package (rt2500). You also need some of the wireless tools from the *wireless-tools* package.

I use this wireless network for four main reasons:

1. **Development.** It is much easier to develop in a comfortable chair with a large monitor than to do it directly with Groucho's monitor.
2. **Communication.** I can monitor Groucho's progress.
3. **Direct control.** I can control Groucho via the Web and move him about the house. This was very useful in getting test images and sensor data, so that I could set the programming parameters.
4. **Spying.** By checking the image from Groucho's camera, I can find out what my dog is doing. Unfortunately, this didn't work too well because my dog was avoiding Groucho.

There are many other ways of handling wireless communication. There are many radio modems and radio modules on the market.

WiFi

802.11b and 802.11g are known as WiFi. This is the easiest way to put your robot on the Internet. Linux supports many of the 802.11 chipsets, but it is taking time to get the 802.11g drivers. Luckily, the driver for the RT2500 chipset was available. It is available both from Sourceforge and as a Gentoo package (rt2500). It should be available soon as a standard kernel module.

Of course, by the time you read this, there will be a new generation of wireless cards out there. I'm trying not to think about this, because I don't want to put new cards in all my computers and robots.

The way to check your chipset is, as I've mentioned previously, with the lspci command.

If you can't use a PCI card for some reason, there is another way: Use an Ethernet to the WiFi bridge. I use such a box made by Linksys (Figure 10-1). It costs a bit more than a simple PCI board, but it makes it easy to put a Linux computer on the Internet. It's easier, too.

I admit that I don't have my robot on the real Internet, but I have a small in-house intranet and use that all the time. It's nice to be able to do development on my laptop without the robots being plugged in.

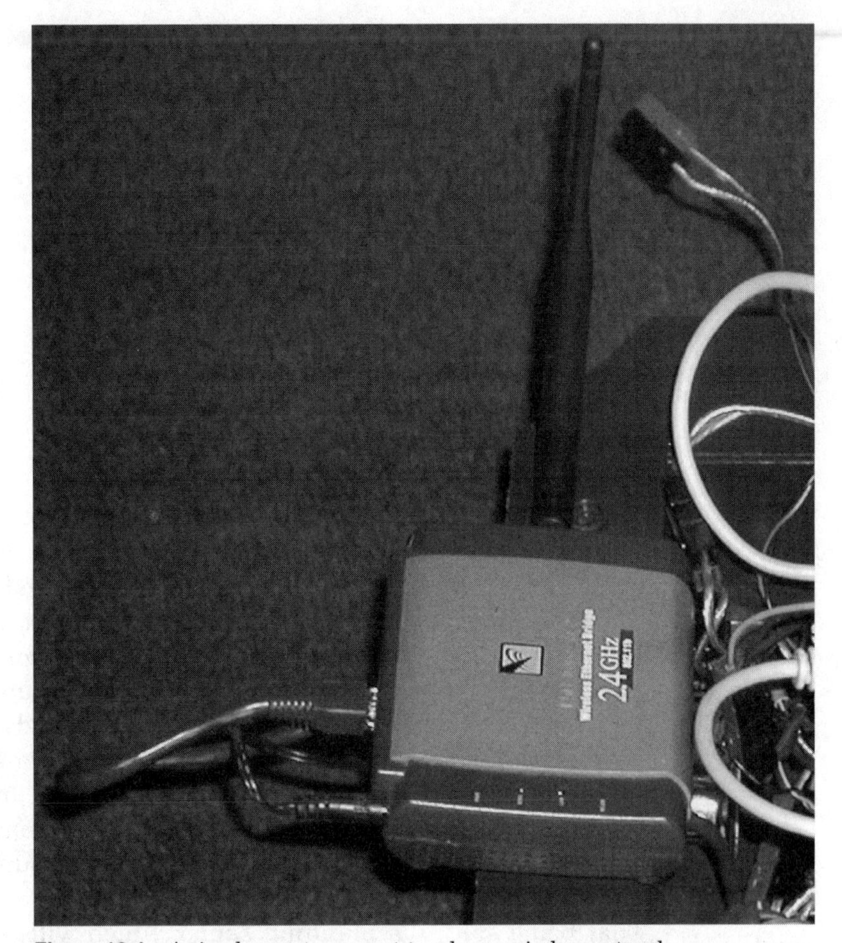

Figure 10-1 A simple way to connect to a home wireless network.

Radio Modems

Another way to get connectivity is to use a radio modem that goes through a serial port. There are a number of manufacturers of such items. I like the products from SparkFun Electronics (http://www.sparkfun.com), which include a number of simple radio modems and Bluetooth modules. You can get similar modules from almost any electronics supplier.

These modules range from those that can be used as easily and reliably as a serial cable to those that just send and receive bytes.

One advantage to using a device like this is that it can have a longer range than a WiFi modem has. Another advantage is that these modules use less power than standard WiFi cards

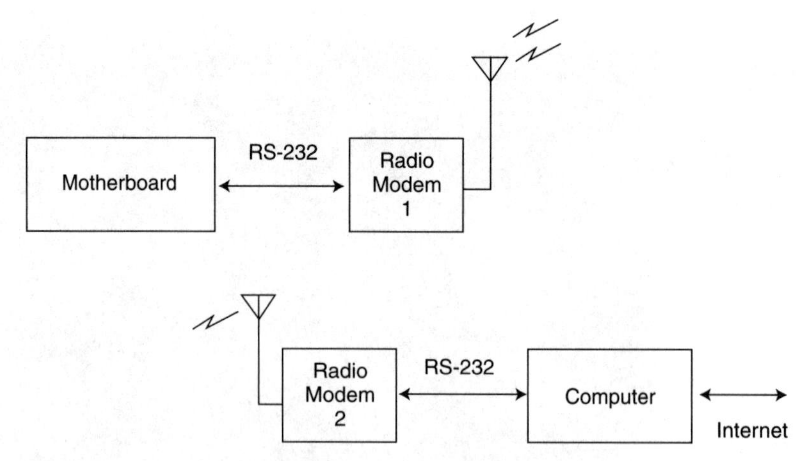

Figure 10-2 Just use the radio modem as you would a serial cable.

do. And you can use PPP or SLIP on it just as you can with a regular serial line (Figure 10-2). You still need another machine to connect the robot to the Internet, if that is desired.

One disadvantage is that you have to do some more work to connect to other computers. However, you can connect to the Internet via a bridge machine. Another minor disadvantage is that the speed might be a bit lower than an 802.11g solution.

The biggest disadvantage, in my opinion, is that by using a nonstandard device, you make it more difficult to connect your robot to another machine. That could be turned around by saying that it makes your robot more secure.

That being said, I'm planning Zeppo, which will be smaller and faster than Groucho, as a mobile sensor platform connected to Groucho via a radio modem (Figure 10-3). Some of the SparkFun devices look interesting, and they're getting better and smaller with time.

It's Connected: Now What?

I think most people who build robots would like to have them connected to the Internet. I would love to have Groucho come in and read me my e-mail. However, that isn't going to happen any time soon. Frankly, I don't want to have my spam read to me. E-mail is one server that I won't be running on any of my robots soon.

There are a number of open-source packages that can be useful on a robot:

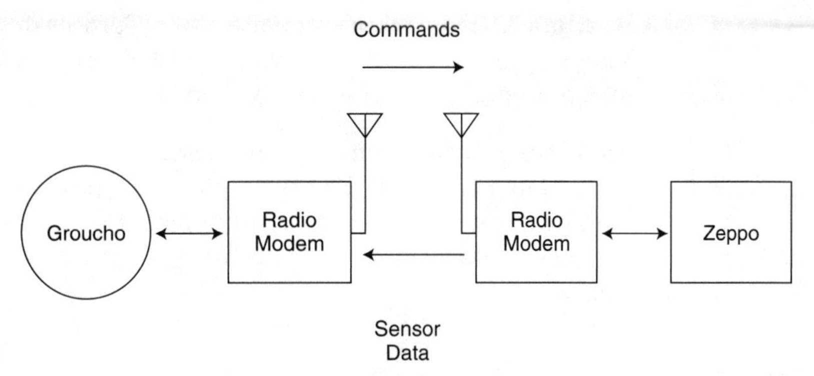

Figure 10-3 A possible way of communicating from robot to robot.

- **The Secure Shell (sshd).** This is a standard Linux/Unix server that allows another computer to log in to your robot securely. It allows me to do much of my programming on my laptop (which runs Windows XP). It replaces telnet for the most part.
- **The Secure File Transfer Protocol.** Most *sshd* servers also include this protocol. It is much more secure than FTP.
- **Telnet (telnet).** This is a basic network terminal program; I do not recommend running telnetd (the telnet server) because the communication is unencrypted. I find the telnet client very useful for testing text-based network protocols such as Festival and Sphinx4.
- **Tomcat.** This is a Web and Java Server Pages (JSP) server. I find it one of the most useful Web servers around.
- **Apache.** This is one of the most popular Web servers.
- **NFS.** The Network File System allows the robot to have a full development environment without using a lot of local disk space
- **Camsource.** This is the V4L image server that Groucho uses. It allows Groucho to access individual frames from any available webcam.
- **PostgreSQL.** This is a very complete SQL database that can work both locally and over the network.
- **Festival.** This is a very powerful text-to-speech processor.
- **Sphinx4.** This is a voice-to-text processor that is written entirely in Java.

- **Sendmail.** This is the generic e-mail-processing server. I have no plan to run it on any of my robots at this time, but if set up properly, it can be fairly secure.

In addition, you can use the network to allow other computers to help the robot. For example, it is possible to set up remote cameras and sensors throughout the house; the robot can check those sensors through the Net, typically through the Web.

Tomcat

This is the Web server I use most often on Groucho. Tomcat is easily installed, either as a Gentoo package (tomcat) or from http://www.apache.org.

I didn't even bother to mess with the configuration files. I just ran it and put my JSP pages in a directory ("Groucho" in my case) under Tomcat's ROOT directory (specifically opt/tomcat5/webapps/ROOT).

One of Tomcat's nicer features is that it serves HTML files as well and can serve older CGI files as well. This means that I don't need to have another Web server on my system.

I like programming network clients as much as the next programmer. However, Web browsers and servers are good enough so that all I have to do is have to program a few HTML pages and JSPs. This saves me a lot of time.

For example, to allow me to control Groucho's actions—in order to get some of the pictures for this book—I merely created a small JSP (Figure 10-4) that I can use via any Web browser, and the results are communicated to Groucho's command processor.

The Java Server Page that does this is fairly basic, sending its commands to Groucho's command processor via port:

```
<%@page import="java.io.*" %>
<%@page import="java.net.*" %>
<%@page import="java.awt.*" %>
<%@page import="java.awt.image.*" %>
<%@page import="javax.swing.ImageIcon" %>
<%@page import="com.sun.image.codec.jpeg.*" %>

<html>
```

Figure 10-4 A command JSP.

```
<head><title>Control Test</title></head>
<body>
<h1>Groucho</h1>

<%
   String s = request.getParameter("s");
   String data = null;
%>
   <p>Command = <%= s %>
<%

   if (s != null)  {
            try {
                        Socket socket = new Socket("localhost", 5601);
                        OutputStreamWriter writer =
                                    new
OutputStreamWriter(socket.getOutputStream());
                        writer.write(s + "\n");
                        socket.close();
            }
```

```
                catch (Exception ex) {
                }
    }

    // Create an image
    try {
                // Read the image
                URL url = new URL("http://groucho.home.jay:9192/largequal");
                ImageIcon icon = new ImageIcon(url);
                Image image = icon.getImage();
                BufferedImage bufferedImage =
                            new BufferedImage(image.getWidth(null),
image.getHeight(null),
                                        BufferedImage.TYPE_INT_RGB);
                Graphics2D g = bufferedImage.createGraphics();
                g.drawImage(image, 0, 0, null);

                // Write the image to a jpeg file
                String path =
"/opt/tomcat5/webapps/ROOT/groucho/image.jpg";
                FileOutputStream file = new FileOutputStream(path);
                JPEGImageEncoder encoder =
JPEGCodec.createJPEGEncoder(file);
                encoder.encode(bufferedImage);
                file.flush();
                file.close();
    }
    catch (Exception ex) {
    }


%>

<form method="POST" action="control.jsp" name="f">

<table cellpadding="10" cellspacing="10">
<tr>
    <td colspan="3" align="center">
                <table border="1">
                <tr>
                            <td align="CENTER" colspan="3"><img
src="image.jpg"></td>
```

```
            </tr>
            </table>
    </td>
</tr>
<tr>
    <td> </td>
    <td align="center"><input type="submit" name="s"
value="Forward"></td>
    <td> </td>
</tr>
<tr>
    <td align="center"><input type="submit" name="s" value="Left"></td>
    <td align="center"><input type="submit" name="s" value="Stop"></td>
    <td align="center"><input type="submit" name="s" value="Right"></td>
</tr>
<tr>
    <td> </td>
    <td align="center"><input type="submit" name="s"
value="Reverse"></td>
    <td> </td>
</tr>
</table>

</form>
</body>
</html>
```

Apache

Apache is one of the most popular, perhaps the most popular, Web servers. Setup is extremely easy. The server is extremely powerful.

For a robot, though, I would rather run one of the many smaller Web servers out there. My robots don't need a server as powerful as Apache when Tomcat will do just as well. I do like using Apache on a normal server.

However, people are comfortable with different things. If you prefer writing CGIs or PHP to writing JSPs, Apache may work better for you. Basically, everything that I can do with Tomcat I can do with Apache. I've even heard that some people run both Apache and Tomcat together, so that Tomcat handles all the JSPs; I've never been able to do this.

NFS

I used the Network File System (NFS) for a couple of previous Linux robots that had the equivalent only of 1 to 4 gigabytes of hard storage. These robots used Compact Flash (CF) cards instead of hard drives. Using CF cards saves a lot of battery power.

You need a machine with a decent amount of disk space to run an NFS server (*nfsd*). Fortunately for us, this is a very easy thing to do if you have a machine on your home network with some extra disk space.

I mounted the following file systems on the NFS server:

```
/usr/src
/usr/portage
/var/tmp/portage
```

The configuration file I use on my host machine is in /etc/exports:

```
/usr/gentoo-src stuart.home.jay(rw,no_subtree_check,sync,no_root_squash)
/usr/gentoo-portage
stuart.home.jay(rw,no_subtree_check,sync,no_root_squash)
/usr/gentoo-portage-tmp
stuart.home.jay(rw,no_subtree_check,sync,no_root_squash)
```

This configuration file allows one of my robots (Stuart) to access the NFS file system only for the three specific files. Stuart's fstab file defines how they are accessed. Stuart is set up so that I have to mount these file systems specifically. This was done because I didn't want to mount them unless I was connected to my home network.

By doing this, I can recompile my robot's system without having to use a cross-compiler *and* without having to use a large hard drive. Since I wouldn't want to use these file systems unless I was doing a recompile, I don't lose anything.

I find that this works fine for robotics.

Yes, using the network makes compiling things a bit slower, but I prefer compiling system software on the same system that runs it.

NFS isn't the only file system that works over the network. There are also the Andrew File System (AFS), Coda, and Samba, among many others.

Camsource

Camsource is a very nice V4L (Video for Linux) image server. It takes any video source, such as a webcam, with a V4L driver and serves either streaming video or individual frames. A nice feature is that several filters are included, so that some preprocessing of the image can be done by Camsource.

The images are served as JPEG images through an HTTP connection on port 9192.

There is a self-descriptive XML setup file that describes the action of Camsource. This file sets up the image size, number of frames, and filters used. It is possible to serve multiple versions of an image. The filter that I use is one that resizes the 160-by-120 frame to 80-by-60. The smaller frame is much easier to deal with.

The parts of the configuration file (/etc/camsource.conf) that I use are:

```
<camsourceconfig>
  <!-- Global config options -->
  <camdev>
        <!--
              This section gives config options for the
              camera device.
        -->

        <plugin>input_v4l</plugin>

        <path>/dev/video</path>
        <width>160</width>
        <height>120</height>
        <fps>10</fps>
        <channel>0</channel>
        <norm>auto</norm>

        <brightness>32767</brightness>
        <hue>32767</hue>
        <colour>32767</colour>
        <contrast>32767</contrast>
        <whiteness>32767</whiteness>
  </camdev>
  <!--
```

```
                The http module. It lets you create virtual paths to
                serve various variants of the current image.
    -->
    <module name="http" active="yes">
                <port>9192</port>
                <vpath>
                            <path>/</path>
                            <path>/small</path>
                            <filter name="resize">
                                        <scale>50</scale>
                            </filter>
                </vpath>
    </module>
</camsourceconfig>
```

A couple of code snippets are used to get a picture from Camsource and into a Java application (Figure 10-5). The basic assumption is that the URL for the image is at http://local host:9192.

Figure 10-5 Picture: chair.

```
// Get an Image the easy way
public static Image getImage(URL url) {
    ImageIcon icon = new ImageIcon(url);
    return icon.getImage();
}
// Draw an Image into a BufferedImage
public static BufferedImage getBufferedImage(Image image) {
    BufferedImage b = new BufferedImage(image.getWidth(null),
            image.getHeight(null), BufferedImage.TYPE_INT_RGB);
    Graphics2D g = b.createGraphics();
    g.drawImage(image, 0, 0, null);
    return b;
}
// Call the two methods with one convenience method
public static BufferedImage getBufferedImage(URL url) {
    return getBufferedImage(getImage(url));
}
```

This code is taken from ws.enerd.robots.image.ImageMinder. To get an image from Camsource, you just need a few lines:

```
import ws.enerd.robots.image.ImageMinder;
BufferedImage image =
ImageMinder.getBufferedImage(
new URL("http://localhost:9192/")
);
```

The default URL that Camsource uses is 9192, and it returns a JPEG image.

PostgreSQL

Although PostgreSQL can be run over a network, I generally use it as a local database. Currently, I'm not using it for anything, but I'm thinking about using it to store map and image data.

Assuming that the map was stored in the database, I could send an SQL statement to Groucho from another computer to return the map or a route. I've been thinking about how I could build an occupancy map using an SQL database. It then should be possible to localize by using the immediate data to get a list of possible locations with a single SQL statement.

Some people prefer MySQL, but I prefer PostgreSQL. This may be just personal preference.

The advantages of using a powerful database include the following:

- Allowing a database to hold the data you wish to save between reboots
- The ability to share the data over the network
- Letting a database do what it does best (store and retrieve data)

Code

For communicating from these servers to Groucho's central code, I generally use the StringSensor class in order to accept data:

```
public class StringSensor extends AbstractSensor {
   public StringSensor(SensorImpl newSensorImpl, int newReadInterval) {
            super(newSensorImpl, 0, newReadInterval);
   }
   public void setValue(int newValue) {
            sensorValue = newValue;
   }
   public void setObjectValue(Object o)  {
            if (o != objectValue) {
                     objectValue = o;
                     setChanged(true);
            }
   }
   public String getString() {
            String s = (String)objectValue;
            objectValue = null;
            return s;
   }
}
```

I choose to use the Sensor interface in order to use the basics of behavioral programming along with other sensors.

SocketSensorImpl

The SocketSensorImpl was designed so that Groucho's central code opens a socket to another server and reads data when available. I use this SensorImpl when I want to communicate with the Sphinx4 service:

```
public class SocketSensorImpl extends AbstractSensorImpl {
   protected String host = null;
   protected int port = 0;
   protected Socket socket = null;
   protected BufferedReader reader = null;
   public SocketSensorImpl(String h, int p) {
            this.host = h;
            this.port = p;
            openSocket();
   }
   public void openSocket() {
            if ((socket != null) && socket.isConnected()) {
                     return;
            }
            // Now open the socket
            try {
                     socket = new Socket(host, port);
                     reader = new BufferedReader(new
InputStreamReader(socket
                                              .getInputStream()));
            } catch (UnknownHostException e) {
                     e.printStackTrace();
            } catch (IOException e) {
                     e.printStackTrace();
            }
   }
   public void readSensor() {
            String s = null;
            openSocket();
            try {
                     if (reader.ready()) {
                              s = reader.readLine();
                     }
                     if (s != null) {
                              s.trim();
                     }
                     objectValue = s;
                     intValue = ((s == null) || (s.length() == 0)) ? 0 : 1;
            } catch (IOException e) {
```

```
                e.printStackTrace();
            }
        }
    }
```

This code merely opens a socket and checks to see if any data are available. If the data are available, the SensorListener is called. If the socket happens to close, it is reopened.

ServerSocketSensorImpl

The ServerSocketSensorImpl was designed to create a simple server and read data. This class was created specifically so that I could use a JSP or CGI to control Groucho.

```
public class SocketServerSensorImpl extends AbstractSensorImpl {
    class ServerThread extends Thread {
        ServerSocket socket = null;
        Vector lines = new Vector();
        public ServerThread(int p) {
            try {
                socket = new ServerSocket(p);
            } catch (IOException ex) {
                ex.printStackTrace();
            }
        }
        public void run() {
            while (true) {
                try {
                    Socket client = socket.accept();
                    BufferedReader reader = new
BufferedReader(
                            new
InputStreamReader(client.getInputStream()));
                    while (!client.isClosed()) {
                        String s = reader.readLine();
                    }
                } catch (IOException ex) {
                    ex.printStackTrace();
                }
                Thread.yield();
            }
```

```
            }
            public String getFirstLine() {
                String s = null;
                synchronized (lines) {
                    if (lines.size() > 0) {
                        s = (String) lines.get(0);
                        lines.removeElementAt(0);
                    }
                }
                return s;
            }
        }
        protected ServerThread thread = null;
        public SocketServerSensorImpl(int p) {
            thread = new ServerThread(p);
        }
        public void readSensor() {
            String s = thread.getFirstLine();
            objectValue = s;
            if (s == null) {
                intValue = 0;
            } else {
                intValue = 1;
            }
        }
}
```

This sensor is used only for testing and remote control.

Summary

- Like any Linux machine, a Linux-based robot can run standard network services.
- For a mobile robot, it is best to use wireless communications.
- The two main ways of using wireless are WiFi and serial radio modules.
- In order to use some text-based services, I use the Sensor interface.

Appendix

ENerd Robotics Framework Code

Main Interfaces

These interfaces are the main interfaces that form the ENerd Robotics Framework.

ws.enerd.robots.Motor.java

```
/*
 * Created on Mar 20, 2005
 *
 * Motor.java - An interfact to a generic motor
 * Copyright 2005 by D. Jay Newman; All rights reserved
 *
 * The licence is the Attribution-ShareAlike 2.0 License as stated at:
 * http://creativecommons.org/licenses/by-sa/2.0/
 *
 */
package ws.enerd.robots;

/**
 * @author jay
 *
 * <p>
 * An interface to a generic motor
 *
 * <p>
 * The licence is the Attribution-ShareAlike 2.0 License as stated at:
```

```
 * http://creativecommons.org/licenses/by-sa/2.0/
 */
public interface Motor {
    public static final Object FORWARD = new Object();
    public static final Object BACKWARD = new Object();

    // Motor initialization
    public void initMotor();
    public void closeMotor();

    // The current or target power
    public void setPower(int newPower);
    public int getPower();

    // The maximum power level for this motor
    public void setMaxPower(int newMaxPower);
    public int getMaxPower();

    // Keep the motor going with the same power, but in the
    // opposite direction; this can be dangerous with some
    // motors and geartrains
    public void reverse();

    // Set/get the current direction
    public void setDirection(Object newDirection);
    public Object getDirection();

    // Whatever the speed, put the motor into dynamic braking
    // mode and set the power to 0; if a motor doesn't support
    // dynamic braking, then just set the power to 0.
    public void brake();

    // Optionally supported commands

    // Return the current actual speed of the motor
    public int getActualSpeed();

    // These concern the position of the motor which should
    // relate linearly to distance
    public void resetPosition();
    public int getPosition();
```

```
    // PID Paramters
    public void setPID(int periodMs, double p, double i, double d);
}
```

ws.enerd.robots.Sensor.java

```
/*
 * Created on Mar 23, 2005
 *
 * The licence is the Attribution-ShareAlike 2.0 License as stated at:
 * http://creativecommons.org/licenses/by-sa/2.0/
 */

/**
 * Sensor.java
 *
 * <p>
 * An interface to a generic sensor
 *
 * <p>
 * The sensor uses a SensorImpl implementation to hold the actual
 * implementation of the physical part. This way I can use the same
 * Sensor implementation for the same type of sensor on different systems.
 *
 * <p>
 * The licence is the Attribution-ShareAlike 2.0 License as stated at:
 * http://creativecommons.org/licenses/by-sa/2.0/
 *
 * @author D. Jay Newman
 */

package ws.enerd.robots;

import ws.enerd.robots.sensor.SensorImpl;
import ws.enerd.robots.sensor.SensorListener;
import ws.enerd.robots.sensor.SensorFilter;

public interface Sensor {
    // Return the main value of the sensor as an integer or object
    public int getValue();
    public Object getObjectValue();
```

```
        // The interval in milliseconds between reads
        public void setReadInterval(int newReadInterval);

        public int getReadInterval();

        // An active sensor is somehow broadcasting; some sensors need to be
        // activated before having their value read
        public void setActive(boolean active);

        public boolean isActive();

        // Handle the case of a filtered sensor
        public void addSensorFilter(SensorFilter newFilter);

        public void removeSensorFilter(SensorFilter newFilter);

        public void filterSensor();

        // Deal with the SensorListener's
        public void addSensorListener(SensorListener listener);

        public void removeSensorListener(SensorListener listener);

        // Set the value and if changed call fireSensorListeners
        public void setValue(int newValue);

        // Some sensors have values that can't be reduced to an integer
        public void setObjectValue(Object o);

        public void fireSensorListeners();

        public boolean shouldRead(long time);

        public void readSensor(long time);

        public void setSensorImpl(SensorImpl impl);
    }
```

ws.enerd.robots.Robot.java

```
    /*
     * Created on Mar 30, 2005
```

```
 *
 * The licence is the Attribution-ShareAlike 2.0 License as stated at:
 * http://creativecommons.org/licenses/by-sa/2.0/
 */
package ws.enerd.robots;

import ws.enerd.robots.behavior.*;
import ws.enerd.robots.sensor.*;

/**
 * ROBOT
 *
 * <p>
 * This is a generic interface to the central processing of your robot.
 *
 * <p>
 * A robot can extend this class to provide the main services that a robot
 * needs.
 *
 * <p>
 * Copyright 2005 D. Jay Newman
 *
 * <p>
 * The licence is the Attribution-ShareAlike 2.0 License as stated at:
 * http://creativecommons.org/licenses/by-sa/2.0/
 *
 * @author D. Jay Newman
 */
public abstract class Robot extends Thread implements SensorListener {
    protected Arbitrator arbitrator = null;
    protected SensorThread sensorThread = null;

    /**
     * initRobot
     *
     * <p>
     * Setup the main objects for the robot; this should be
     * called to initialize the robot.
     *
     * @param a The Arbitrator to use for this robot
     */
    protected void initRobot() {
```

```
                makeArbitrator();
                makeSensorThread();
                makeConnections();
                makeSensors();
                makeActuators();
                makeBehaviors();
}

/**
 * startRobot
 *
 * <p>
 * Call this to start the robot
 *
 */
protected void startRobot() {
                arbitrator.startArbitrator();
                sensorThread.start();
}

/**
 * makeArbitrator
 *
 * <p>
 * Create the Arbitrator for this robot
 *
 */
protected void makeArbitrator() {
}

/**
 * getArbitrator
 *
 * <p>
 * Get the Arbitrator if you need access
 *
 * @return The robot's Arbitrator
 */
public Arbitrator getArbitrator() {
                return arbitrator;
}
```

```
/**
 * makeSensorThread
 *
 * <p>
 * Create the SensorThread for the robot
 */
protected void makeSensorThread() {
        sensorThread = new SensorThread();
}

/**
 * makeSensors
 *
 * <p>
 * Create the actual sensors
 */
protected void makeSensors() {
}

/**
 * makeConnections
 *
 * <p>
 * Any network connections should be created here
 */
protected void makeConnections() {
}

/**
 * makeBehaviors
 *
 * Create behaviors here
 */
protected void makeBehaviors() {
}

/**
 * makeActuators
 *
 * <p>
 * Create your motor objects here
 */
```

```
protected void makeActuators() {
}

/**
 * addSensor
 *
 * <p>
 * Add a sensor to the SensorThread
 *
 * @param s The sensor to add
 */
public void addSensor(Sensor s) {
        sensorThread.addSensor(s);
}

/**
 * removeSensor
 *
 * <p>
 * Remove a sensor from the SensorThread
 *
 * @param s The sensor to be removed
 */
public void removeSensor(Sensor s)  {
        sensorThread.removeSensor(s);
}

/**
 * addBehavior
 *
 * <p>
 * Add a behavior to the arbitrator; this is just
 * a convinience method
 *
 * @param b The Behavior to add
 */
public void addBehavior(Behavior b) {
        arbitrator.addBehavior(b);
}

/**
 * removeBehavior
```

```
     *
     * <p>
     * Remove a behavior from the Arbitrator
     *
     * @param b The Behavior to be removed
     */
    public void removeBehavior(Behavior b) {
            arbitrator.removeBehavior(b);
    }

    /**
     * sensorChanged
     *
     * @see
ws.enerd.robots.sensor.SensorListener#sensorChanged(ws.enerd.robots.s
ensor.SensorEvent)
     */
    public void sensorChanged(SensorEvent e) {
            // TODO Auto-generated method stub

    }

}
```

ws.enerd.robots.DriveSystem.java

```
/*
 * Created on Apr 26, 2005
 *
 * The licence is the Attribution-ShareAlike 2.0 License as stated at:
 * http://creativecommons.org/licenses/by-sa/2.0/
 */
package ws.enerd.robots;


/**
 * DriveSystem
 *
 * <p>
 * An abstraction of a system used to drive and steer a robot.
 *
 * <p>
```

```
 * Currently this is incomplete
 *
 * <p>
 * Copyright 2005 D. Jay Newman
 *
 * <p>
 * The licence is the Attribution-ShareAlike 2.0 License as stated at:
 * http://creativecommons.org/licenses/by-sa/2.0/
 *
 * @author D. Jay Newman
 */
public interface DriveSystem {
   // Informational queries
   public boolean canTurnInPlace();
   public boolean canTellPosition();

   // Information
   public int getMaxSpeed();


   // Setting important variables
   public void setMaxSpeed(int maxSpeed);

   public void forward(int speed);
   public void backward(int speed);

   // OK, this is pretending that the driving is by joystick
   // where speed is from 0 to maxSpeed and angle is 0 - 360
   // (0 being straight ahead
   // degrees.
   //
   // The cos(angle) is the drive; the sin(angle) is the turning.
   public void drive(int speed, int angle);

   // OK, now the speed is from 0.0 - 1.0 and the direction is
   // from -PI to PI with 0 being straight ahead.
   public void drive(double speed, double angle);

   public void reverse();
   public void brake();
   public void brakeNOW();
}
```

Utility

The StoppableThread was created to let me easily create a Thread which can be stopped safely.

ws.enerd.util.StoppableThread.java

```
/*
 * StoppableThread.java
 *
 * Created on Mar 28, 2005
 *
 * The licence is the Attribution-ShareAlike 2.0 License as stated at:
 * http://creativecommons.org/licenses/by-sa/2.0/
 */
package ws.enerd.util;

/**
 * StoppableThread
 *
 * <p>
 * A Thread object which can be stopped by a boolean because I got tired
of
 * writing this into every thread and Runnable I created.
 *
 * <p>
 * Modify the method doRun() to put the running object in it and make sure
 * that doRun() returns on a regular basis.
 *
 * <p>
 * The licence is the Attribution-ShareAlike 2.0 License as stated at:
 * http://creativecommons.org/licenses/by-sa/2.0/
 *
 * @author D. Jay Newman
 */
abstract public class StoppableThread extends Thread {
    protected boolean running = true;
    protected boolean stopped = false;

    public StoppableThread() {
    }
```

```
/**
 * shouldStop
 *
 * <p>
 * Let the thread know that it should stop
 */
public void shouldStop() {
        running = false;
}

/**
 * isRunning
 *
 * <p>
 * Return true as long as the thread is running
 *
 * @return
 */
public boolean isRunning() {
        return running;
}

/**
 * isStopped
 *
 * <p>
 * Return true when the thread is stopped
 *
 * @return
 */
public boolean isStopped()  {
        return stopped;
}

/**
 * run
 *
 * <p>
 * Basically keep calling the doRun method until the running
 * is set to false
 */
public void run() {
```

```
                running = true;
                stopped = false;
                while (running) {
                        doRun();
                        Thread.yield();
                }

                stopped = true;
        }

        /**
         * doRun
         *
         * <p>
         * This method must be overridden and is called repeatedly
         * until told to stop
         */
        abstract public void doRun();
}
```

MojoBus

The MojoBus network is used to communicate with the BDMicro RX50 prototype motor controllers as well as the MAVRIC-IIb which is used as a sensor processor.

This is a basic text-based master-slave protocol. I am running this over an RS-485 connection, but it can be run over almost any connection that allows for multiple nodes.

ws.enerd.robots.mojo.MojoConstants.java

```
/*
 * Created on May 15, 2005
 *
 * The licence is the Attribution-ShareAlike 2.0 License as stated at:
 * http://creativecommons.org/licenses/by-sa/2.0/
 */

/*
 * Mojotalk:
 *
 * Packet format:
```

```
 *
 *   [>]>t[,f]:cmd;[cmd;cmd;...]
 *
 * Where:
 *
 *   t   = bus id of recipient (required)
 *
 *   f   = bus id of sender (optional)
 *
 *   cmd = command to execute
 *
 * Command format:
 *
 *   parameter | parameter=value
 *
 *
 * Parameters:
 *
 *   RESET   = ro, reset system
 *   VERSION = ro, display firmware revision
 *   ID      = rw, set controller bus ID
 *   BAUD    = rw, set RS232/RS485 serial baud rate, units = bps
 *   VERBOSE = rw, set verbosity
 *   ECHO    = rw, set input echo
 *
 */

package ws.enerd.robots.mojo;

/**
 * MojoConstants
 *
 * <p>
 * Copyright 2005 D. Jay Newman
 *
 * <p>
 * The licence is the Attribution-ShareAlike 2.0 License as stated at:
 * http://creativecommons.org/licenses/by-sa/2.0/
 *
 * @author D. Jay Newman
 */
public interface MojoConstants {
```

```
    // Commands
    public final String MOJO_RESET = "RESET";
    public final String MOJO_VERSION = "VERSION";
    public final String MOJO_ID = "ID";
    public final String MOJO_BAUD = "BAUD";
    public final String MOJO_SAVE_BAUD = "SAVEBAUD";
    public final String MOJO_VERBOSE = "VERBOSE";
    public final String MOJO_ANNOUNCE = "ANNC";
    public final String MOJO_WHO = "WHO";

    // Addresses
    public final int MOJO_ADDR_ALL = 0;
    public final int MOJO_ADDR_NONE = -1;
}
```

ws.enerd.robots.mojo.MojoDevice.java

```
/*
 * Created on May 23, 2005
 *
 * The licence is the Attribution-ShareAlike 2.0 License as stated at:
 * http://creativecommons.org/licenses/by-sa/2.0/
 */
package ws.enerd.robots.mojo;

/**
 * MojoDevice
 *
 * <p>
 * This interface should be used by all devices that use
 * a MojoConnection
 *
 * <p>
 * It's a neat idea, but I don't use it.
 *
 * <p>
 * Copyright 2005 D. Jay Newman
 *
 * <p>
 * The licence is the Attribution-ShareAlike 2.0 License as stated at:
 * http://creativecommons.org/licenses/by-sa/2.0/
 *
```

```
 * @author D. Jay Newman
 */

public interface MojoDevice {
    public final int BAUD_9600 = 9600;
    public final int BAUD_115K = 115200;
    public final int BAUD_230K = 230400;

    public void setBaudRate(int baud);
    public void setId(int id);
}
```

ws.enerd.robots.mojo.MojoCommand.java

```
/*
 * Created on May 15, 2005
 *
 * The licence is the Attribution-ShareAlike 2.0 License as stated at:
 * http://creativecommons.org/licenses/by-sa/2.0/
 */
package ws.enerd.robots.mojo;

/**
 * MojoCommand
 *
 * <p>A command for the MojoBus protocol
 *
 * <p>
 * Copyright 2005 D. Jay Newman
 *
 * <p>
 * The licence is the Attribution-ShareAlike 2.0 License as stated at:
 * http://creativecommons.org/licenses/by-sa/2.0/
 *
 * @author D. Jay Newman
 */
public class MojoCommand implements MojoConstants {

    public static void main(String[] args) {
    }

    private String command = null;
```

```
    private String parameter = null;
    private int destination = 0;

    public MojoCommand(String command, String parameter, int destina-
tion) {
            this.command = command;
            this.parameter = parameter;
            this.destination = destination;
    }

    public MojoCommand(String command, int destination) {
            this(command, null, destination);
    }

    public MojoCommand(String command) {
            this(command, null, MOJO_ADDR_ALL);
    }

    public String toString() {
            String s = ">>" + destination + ":" + command;
            if (parameter != null) {
                    s += "=" + parameter;
            }

            s += ";";

            return s;
    }
}
```

ws.enerd.robots.mojo.MojoConnection.java

```
/*
 * Created on Apr 29, 2005
 *
 * The licence is the Attribution-ShareAlike 2.0 License as stated at:
 * http://creativecommons.org/licenses/by-sa/2.0/
 */
package ws.enerd.robots.mojo;

import java.io.*;
```

```
/**
 * MojoConnection
 *
 * <p>
 * This is a serial connection that allows one to write commands and read
 * responses via streams that talk MojoTalk (actually this will talk with
 * anything with both an InputStream and an OutputStream.
 *
 * <p>
 * Commands and responses are sent over the network in MojoTalk, but
 * communication to and from this connection to Java is via the
MojoCommand and
 * MojoResponse object.
 *
 * <p>
 * This connection runs on the honor system. Any calling object must
 * wait for the Lock to be free and then Lock and Release it on their
 * own.
 *
 * <p>
 * Copyright 2005 D. Jay Newman
 *
 * <p>
 * The licence is the Attribution-ShareAlike 2.0 License as stated at:
 * http://creativecommons.org/licenses/by-sa/2.0/
 *
 * @author D. Jay Newman
 */
public class MojoConnection {
    private BufferedReader reader = null;
    private BufferedWriter writer = null;

    private boolean lock = false;

    /**
     * Constructor
     *
     * <p>
     * Given an InputStream and an OutputStream, create the
     * proper Reader and Writer.
     *
     * @param in
```

```
* @param out
*/
public MojoConnection(InputStream in, OutputStream out) {
        reader = new BufferedReader(new InputStreamReader(in));
        writer = new BufferedWriter(new OutputStreamWriter(out));

        try {
                while (reader.ready()) {
                        int i = reader.read();
                }
        }
        catch (Exception ex) {
                ex.printStackTrace();
        }
}

public String readLine() throws IOException {
        String s = reader.readLine();
        while ((s == null) || (s.trim().length() == 0)) {
                s = reader.readLine();
        }
        return s;
}

public void writeLine(String line) throws IOException {
        writer.write(line);
        writer.newLine();
        writer.flush();
}

public boolean ready() throws IOException {
        return reader.ready();
}

public synchronized boolean getLock() {
        if (lock == false) {
                lock = true;
                return true;
        }

        return false;
}
```

```
    public void writeCommand(MojoCommand command) throws
IOException {
            writeLine(command.toString());
    }

    public void waitForLock() {
            while (!getLock()) {
                    Thread.yield();
            }
    }

    public synchronized void releaseLock() {
            lock = false;
    }
}
```

ws.enerd.robots.mojo.TestMotors.java

```
/*
 * Created on May 23, 2005
 *
 * The licence is the Attribution-ShareAlike 2.0 License as stated at:
 * http://creativecommons.org/licenses/by-sa/2.0/
 */
package ws.enerd.robots.mojo;

import java.io.*;

import gnu.io.*;

import ws.enerd.robots.motor.*;

/**
 * TestMotors
 *
 * <p>
 * Copyright 2005 D. Jay Newman
 *
 * <p>
 * The licence is the Attribution-ShareAlike 2.0 License as stated at:
 * http://creativecommons.org/licenses/by-sa/2.0/
```

```
 *
 * @author D. Jay Newman
 */
public class TestMotors {

    public static void main(String[] args) {
            InputStream in = null;
            OutputStream out = null;

            // Create the connection
            RXTXCommDriver driver = new RXTXCommDriver();
            SerialPort port = (SerialPort)
driver.getCommPort("/dev/tts/USB0",
                        CommPortIdentifier.PORT_SERIAL);
            try {
                  port.setSerialPortParams(9600 /* 115200 */,
SerialPort.DATABITS_8,
                              SerialPort.STOPBITS_1,
SerialPort.PARITY_NONE);
                  in = port.getInputStream();
                  out = port.getOutputStream();
            } catch (Exception ex) {
                  ex.printStackTrace();
            }

            MojoConnection connection = new MojoConnection(in, out);
            System.out.println("connection = " + connection);
            MojoMotor left = new MojoMotor(connection, 1);
            MojoMotor right = new MojoMotor(connection, 2);
            right.reverse();

            left.setPower(50);
            right.setPower(50);

            try {
                  Thread.sleep(10 * 1000);
            }
            catch (Exception ex) {}

            left.setPower(0);
            right.setPower(0);
```

```
                System.exit(0);
    }
}
```

Motors

Control of motors is an important part of robotics. Groucho uses the MojoMotor to control the RX50 motor controllers. The motors have high resolution encoders on them which is used by the MojoMotor class.

It is possible to run motors without encoders, and I have done so in many cases. Originally I wrote MojoMotor to handle motors without encoders but it didn't make sense after I had hooked up the encoders.

ws.enerd.robots.motor.AbstractMotor.java

```
/*
 * Created on Mar 21, 2005
 *
 * The licence is the Attribution-ShareAlike 2.0 License as stated at:
 * http://creativecommons.org/licenses/by-sa/2.0/
 */
package ws.enerd.robots.motor;

import ws.enerd.robots.Motor;

/**
 * AbstractMotor.java
 *
 * <p>
 * A basic implementation of the Motor Interface
 *
 * @author D. Jay Newman
 *
 * <p>
 * The licence is the Attribution-ShareAlike 2.0 License as stated at:
 * http://creativecommons.org/licenses/by-sa/2.0/
 */
public abstract class AbstractMotor implements Motor {
    protected int power = 0;
    protected int maxPower = 100;
```

```
protected Object direction = FORWARD;

private AbstractMotor() {
}

/**
 * Constructor
 */
public AbstractMotor(int startingMaxPower) {
        initMotor();
        setMaxPower(startingMaxPower);
}

/**
 * initMotor
 *
 * <p>
 * Anything necessary to initialize the motor
 */
public void initMotor() {
}

/**
 * closeMotor
 *
 * <p>
 * Anything necessary to do to shut-down this driver
 */
public void closeMotor() {
}

/**
 * setPower
 *
 * <p>
 * Set the motor's power (torque, PWM, whatever)
 */
public void setPower(int newPower) {
        power = newPower;
        if (newPower >= 0) {
                direction = FORWARD;
        } else {
```

```
                direction = BACKWARD;
                power = -power;
        }

        if (power > maxPower) {
                power = maxPower;
        }

        applyMotorPower();
}

/**
 * getPower
 *
 * <p>
 * Get the motor's power. If possible, return the
 * actual power.
 */
public int getPower() {
        return power;
}

/**
 * setMaxPower
 *
 * <p>
 * Set the maximum power for the motor. This is
 * necessary for some equations to work out.
 */
public void setMaxPower(int newMaxPower) {
        maxPower = newMaxPower;
}

/**
 * getMaxPower
 *
 * <p>
 * Return the previously set maxPower
 */
public int getMaxPower() {
        return maxPower;
}
```

```
/**
 * reverse
 *
 * <p>
 * If possible reverse the direction of the motor. The default
 * method does nothing because reversing some motor/controler
 * combinations can cause massive power spikes.
 */
public void reverse() {
}

/**
 * setDirection
 *
 * <p>
 * OK, I know what I said in the "reverse" method, but if
 * you really want to do a reverse, that's your problem.
 */
public void setDirection(Object newDirection) {
        if ((direction != FORWARD) && (direction != BACKWARD)) {
                direction = FORWARD;
        }

        direction = newDirection;
        applyMotorPower();
}

/**
 * getDirection
 *
 * <p>
 * Return the direction of the motor
 */
public Object getDirection() {
        return direction;
}

/**
 * brake
 *
 * <p>
```

```
         * If possible, use dynamic braking; the default implementation
         * merely sets the motor power to 0.
         */
        public void brake() {
                power = 0;
                applyMotorPower();
        }

        /**
         * applyMotorPower
         *
         * <p>
         * This is the only method that absolutely needs to be
         * overridden.
         */
        abstract protected void applyMotorPower();

        // Optionally supported commands

        // You need odometry for this
        public int getActualSpeed() {
                return getPower();
        }

        public void resetPosition() {
        }

        public int getPosition() {
                return 0;
        }

        // PID Paramters
        public void setPID(int periodMs, double p, double i, double d) {
        }
}
```

ws.enerd.robots.motor.MojoMotor.java

```
/*
 * Created on Apr 29, 2005
 *
 * The licence is the Attribution-ShareAlike 2.0 License as stated at:
```

```
 * http://creativecommons.org/licenses/by-sa/2.0/
 */
package ws.enerd.robots.motor;

import ws.enerd.robots.mojo.*;

/**
 * MojoMotor
 *
 * <p>
 * This is a motor that is controlled by a BDMicro RX50 motor
controller<br>
 * http://www.bdmicro.com
 *
 * <p>
 * Copyright 2005 D. Jay Newman
 *
 * <p>
 * The licence is the Attribution-ShareAlike 2.0 License as stated at:
 * http://creativecommons.org/licenses/by-sa/2.0/
 *
 * @author D. Jay Newman
 */
public class MojoMotor extends AbstractMotor implements MojoDevice {
    protected MojoConnection connection = null;
    protected int mojoId = 0;
    protected int savedPosition = 0;

    /**
     * The constructor that uses PID
     *
     * @param con MojoConnection
     * @param mojoId The int id of the device
     * @param encp The time unit for the PID equation
     * @param kp The Kp parameter for the PID equation
     */
    public MojoMotor(MojoConnection con, int mojoId, int encp, double kp) {
            super(1000);
            connection = con;
            setId(mojoId);
            setPID(encp, kp, 0, 0);
            resetPosition();
```

```
    }

    /**
     * The constructor that doesn't use PID
     * @param con
     * @param mojoId
     */
    public MojoMotor(MojoConnection con, int mojoId) {
            this(con, mojoId, 0, 0);
    }


    /**
     * applyMotorPower
     *
     * <p>
     * Talk to the motor controller to set the power.
     *
     * @see ws.enerd.robots.motor.AbstractMotor#applyMotorPower()
     */
    protected void applyMotorPower() {
            int p = getPower();
            if (getDirection() == BACKWARD)
                    p = -p;
            connection.waitForLock();
            MojoCommand command = new MojoCommand("TVEL", "" +
p, mojoId);
            try {
                    connection.writeCommand(command);
            } catch (Exception ex) {
                    ex.printStackTrace();
            }
            connection.releaseLock();
    }

    /**
     * setBaudRate
     *
     * <p>
     * Set the baud rate for this device
     *
     * @see ws.enerd.robots.mojo.MojoDevice#setBaudRate()
```

```
	 */
	public void setBaudRate(int baud) {
	}

	/**
	 * setId
	 *
	 * <p>
	 * Set the MojoTalk id for this device
	 *
	 * @see ws.enerd.robots.mojo.MojoDevice#setId(int)
	 */
	public void setId(int newId) {
		mojoId = newId;
	}

	/**
	 * reverse
	 *
	 * <p>
	 * Allow this motor to be treated as a resered motor
	 */
	public void reverse() {
		MojoCommand command = new MojoCommand("REV", "1",
mojoId);
		connection.waitForLock();
		try {
			connection.writeCommand(command);
		} catch (Exception ex) {
			ex.printStackTrace();
		}
		connection.releaseLock();
	}

	/**
	 * getActualSpeed
	 *
	 * <p>
	 * Get the actual speed of the motor as measured
	 */
	public int getActualSpeed() {
		int v = 0;
```

```
                String s = "";

                connection.waitForLock();
                try {
                        connection.writeCommand(new
MojoCommand("AVEL", mojoId));
                        while (!s.startsWith("*AVEL=")) {
                                s = connection.readLine();
                        }
                        // Get the numerical part
                        s = s.substring(6, s.length());
                        v = Integer.parseInt(s);
                }
                catch (Exception ex) {
                        ex.printStackTrace();
                }

                return v;
        }

        /**
         * resetPosition
         *
         * <p>
         * Reset the position of the motor
         */
        public void resetPosition() {
                connection.waitForLock();
                try {
                        connection.writeCommand(new
MojoCommand("APOS=0", mojoId));
                }
                catch (Exception ex) {
                        ex.printStackTrace();
                }
        }

        /**
         * getPosition
         *
         * <p>
         * Get the motor's position (distance travelled)
```

```
    */
    public int getPosition() {
            return readPosition();
    }

    /**
     * readPosition
     *
     * <p>
     * Read the position of the motor
     *
     * @return The actual motor position
     */
    protected int readPosition() {
            int pos = 0;
            String s = "";

            connection.waitForLock();
            try {
                    connection.writeCommand(new
MojoCommand("APOS", mojoId));
                    while (!s.startsWith("*APOS=")) {
                            s = connection.readLine();
                    }
                    // Get the numerical part
                    s = s.substring(6, s.length());
                    pos = Integer.parseInt(s);
            }
            catch (Exception ex) {
                    ex.printStackTrace();
            }

            return pos;
    }

    /**
     * setPID
     *
     * <p>
     * Set the PID parameters for this device
     *
     * @param periodMs The time period
```

```
     * @param p The Kp parameter
     * @param i The Ki parameter
     * @param d The Kd parameter
     */
    public void setPID(int periodMs, double p, double i, double d) {
            connection.waitForLock();
            try {
                    connection.writeCommand(new
MojoCommand("ENCP", "" + periodMs, mojoId));
                    connection.writeCommand(new MojoCommand("KP",
"" + p, mojoId));
            }
            catch (Exception ex) {
                    ex.printStackTrace();
            }
            connection.releaseLock();
    }
}
```

ws.enerd.robots.motor.AbstractServoMotor.java

```
/*
 * Created on Mar 20, 2005
 *
 * AbstractServoMotor.java - An abstract class to interface with a
 * normal or converted RC servo motor
 *
 * Copyright 2005 by D. Jay Newman; All rights reserved
 *
 * The licence is the Attribution-ShareAlike 2.0 License as stated at:
 * http://creativecommons.org/licenses/by-sa/2.0/
 */

package ws.enerd.robots.motor;

/**
 * @author D. Jay Newman
 *
 * <p>
 * An interface to a generic R/C Servo.
 *
 * <p>
```

```
 * The licence is the Attribution-ShareAlike 2.0 License as stated at:
 * http://creativecommons.org/licenses/by-sa/2.0/
 */
public abstract class AbstractServoMotor extends AbstractMotor {
   protected double angle = 0.0;

   public AbstractServoMotor(int maxPower) {
            super(maxPower);
   }

   public void setAngle(double newPosition) {
            angle = newPosition;
            positionToPower();
            applyMotorPower();
   }

   // Convert a specific position to a specific power value
   public abstract void positionToPower();
}
```

ws.enerd.robots.motor.AbstractDriveSystem.java

```
/*
 * Created on Jul 10, 2005
 *
 * The licence is the Attribution-ShareAlike 2.0 License as stated at:
 * http://creativecommons.org/licenses/by-sa/2.0/
 */
package ws.enerd.robots.motor;

import ws.enerd.robots.DriveSystem;

/**
 * AbstractDriveSystem
 *
 * <p>
 * An implementation of the basics for a robotic drive system; this
 * is still experimental.
 *
 * <p>
 * Copyright 2005 D. Jay Newman
 *
```

```
 * <p>
 * The licence is the Attribution-ShareAlike 2.0 License as stated at:
 * http://creativecommons.org/licenses/by-sa/2.0/
 *
 * @author D. Jay Newman
 */
public abstract class AbstractDriveSystem implements DriveSystem {

    protected int maxSpeed = 0;

    /**
     *
     */
    public AbstractDriveSystem() {
    }

    // Informational queries
    public abstract boolean canTurnInPlace();
    public abstract boolean canTellPosition();

    /**
     * @see ws.enerd.robots.DriveSystem#maxSpeed()
     */
    public int getMaxSpeed() {
            return maxSpeed;
    }

    public void setMaxSpeed(int i)  {
            maxSpeed = i;
    }

    /**
     * @see ws.enerd.robots.DriveSystem#forward(int)
     */
    public abstract void forward(int speed);

    /* (non-Javadoc)
     * @see ws.enerd.robots.DriveSystem#backward(int)
     */
    public abstract void backward(int speed);

    /* (non-Javadoc)
```

```
	 * @see ws.enerd.robots.DriveSystem#drive(int, int)
	 */
	public void drive(int speed, int direction) {
		double s = (double)speed / (double)maxSpeed;
		double d = (double)direction * Math.PI / 180.0;
		drive(s, d);
	}

	/* (non-Javadoc)
	 * @see ws.enerd.robots.DriveSystem#drive(int, int)
	 */
	public abstract void drive(double speed, double direction);

	/* (non-Javadoc)
	 * @see ws.enerd.robots.DriveSystem#reverse()
	 */
	public abstract void reverse();

	/* (non-Javadoc)
	 * @see ws.enerd.robots.DriveSystem#brake()
	 */
	public void brake() {
		drive(0.0, 0.0);
	}

	/* (non-Javadoc)
	 * @see ws.enerd.robots.DriveSystem#brakeNOW()
	 */
	public void brakeNOW() {
		brake();
	}
}
```

ws.enerd.robots.motor.DifferentialDriveSystem.java

```
/*
 * Created on Jul 10, 2005
 *
 * The licence is the Attribution-ShareAlike 2.0 License as stated at:
 * http://creativecommons.org/licenses/by-sa/2.0/
 */
package ws.enerd.robots.motor;
```

```
import ws.enerd.robots.Motor;

/**
 * DifferenialDriveSystem
 *
 * <p>
 * An implementation used to control a different drive robot; this
 * is experimental and not finished at all.
 *
 * <p>
 * Copyright 2005 D. Jay Newman
 *
 * <p>
 * The licence is the Attribution-ShareAlike 2.0 License as stated at:
 * http://creativecommons.org/licenses/by-sa/2.0/
 *
 * @author D. Jay Newman
 */
public class DifferentialDriveSystem extends AbstractDriveSystem {

    protected Motor left = null;
    protected Motor right = null;

    /**
     *
     */
    public DifferentialDriveSystem(Motor l, Motor r) {
            setMotors(l, r);
            left.setPower(0);
            right.setPower(0);
    }

    /* (non-Javadoc)
     * @see ws.enerd.robots.motor.AbstractDriveSystem#canTurnInPlace()
     */
    public boolean canTurnInPlace() {
            // TODO Auto-generated method stub
            return false;
    }

    /* (non-Javadoc)
```

```
 * @see ws.enerd.robots.motor.AbstractDriveSystem#canTellPosition()
 */
public boolean canTellPosition() {
        // TODO Auto-generated method stub
        return false;
}

/*
 * @see ws.enerd.robots.DriveSystem#forward(int)
 */
public void forward(int speed) {
        left.setPower(speed);
        right.setPower(speed);
}

/*
 * @see ws.enerd.robots.DriveSystem#backward(int)
 */
public void backward(int speed) {
        left.setPower(-speed);
        right.setPower(-speed);
}

/* (non-Javadoc)
 * @see ws.enerd.robots.motor.AbstractDriveSystem#drive(double, dou-
ble)
 */
public void drive(double speed, double direction) {
        // TODO Auto-generated method stub

}

/* (non-Javadoc)
 * @see ws.enerd.robots.DriveSystem#reverse()
 */
public void reverse() {
        // TODO Auto-generated method stub

}

/* (non-Javadoc)
 * @see ws.enerd.robots.DriveSystem#brake()
```

```
     */
    public void brake() {
            left.brake();
            right.brake();
    }

    /**
     * @see ws.enerd.robots.DriveSystem#brakeNOW()
     */
    public void brakeNOW() {
            brake();
    }


    /*
     * @see ws.enerd.robots.DriveSystem#maxSpeed()
     */
    public int getMaxSpeed() {
            return maxSpeed;
    }

    public void setMotors(Motor l, Motor r)  {
            left = l;
            right = r;
    }
}
```

Sensors

The sensor architecture that I have choosen is that of sensors which are each read at specific intervals. The SensorThread handles this period reading.

In order to allow better reuse of code, the Sensor handles the application interface and the SensorImpl handles the hardware sides. This way I can create a DistanceSensor that can handle many different types of sensors.

Sensors based on AbstractSensor use SensorFilters to handle basic filtering operations such as averaging.

Sensors communicate with the rest of the application with SensorEvents and SensorListeners.

ws.enerd.robots.sensor.SensorThread.java

```
/*
 * Created on Mar 28, 2005
 *
 * The licence is the Attribution-ShareAlike 2.0 License as stated at:
 * http://creativecommons.org/licenses/by-sa/2.0/
 */
package ws.enerd.robots.sensor;

// Java standard types
import java.util.*;

// My own types
import ws.enerd.util.StoppableThread;
import ws.enerd.robots.Sensor;

/**
 * SensorThread
 *
 * <p>
 * This is a thread that checks each sensor until the thread is stopped.
There
 * will usually be one main SensorThread associated with each robot.
 *
 * <p>
 * Copyright 2005 D. Jay Newman
 *
 * <p>
 * The licence is the Attribution-ShareAlike 2.0 License as stated at:
 * http://creativecommons.org/licenses/by-sa/2.0/
 *
 * @author D. Jay Newman
 */
public class SensorThread extends StoppableThread {

    private List sensors = new Vector();

    public void addSensor(Sensor sensor) {
            sensors.add(sensor);
    }
```

```
    public void removeSensor(Sensor sensor) {
        sensors.remove(sensor);
    }

    /**
     * doRun
     *
     * <p>
     * Keep running and check each sensor in turn
     */
    public void doRun() {
        Sensor sensor = null;
        long time = System.currentTimeMillis();
        ListIterator iter = sensors.listIterator();

        while (iter.hasNext()) {
            sensor = (Sensor) (iter.next());
            if (sensor.shouldRead(time)) {
                sensor.readSensor(time);
            }
        }
    }
}
```

ws.enerd.robots.sensor.AbstractSensor.java

```
/*
 * Created on Mar 27, 2005
 *
 * The licence is the Attribution-ShareAlike 2.0 License as stated at:
 * http://creativecommons.org/licenses/by-sa/2.0/
 */
package ws.enerd.robots.sensor;

import java.util.Vector;
import java.util.Iterator;
import ws.enerd.robots.Sensor;

/**
 * AbstractSensor
 *
 * <p>
```

```
 * An implementation of the Sensor interface
 *
 * <p>
 * The licence is the Attribution-ShareAlike 2.0 License as stated at:
 * http://creativecommons.org/licenses/by-sa/2.0/
 *
 * @see ws.enerd.robots.Sensor
 * @author D. Jay Newman
 */
public class AbstractSensor implements Sensor {
    // The underlying hardware implementation for this sensor
    protected SensorImpl sensorImpl = null;

    // The current int and Object values for this sensor
    protected int sensorValue = 0;
    protected Object objectValue = null;

    // The default int min and max values for this sensor
    protected int maxValue = Integer.MAX_VALUE;
    protected int minValue = 0;

    // These are private and only handled at this level

    // A SensorFilter: these can be chained to have multiple filters
    private SensorFilter filter = null;

    // The listener list
    private Vector listeners = new Vector();

    // true if the sensor is doing something abnormal
    private boolean active = false;

    // The interval
    private int readIntervalMillis = 100;

    // This is the system time at which the sensor was last read
    private long lastReadTime = System.currentTimeMillis();

    // This is the time for the next sensor reading
    private long nextReadTime = lastReadTime;

    // true if the sensor changed since the last reading
```

```
private boolean sensorChanged = false;

/**
 * AbstractSensor
 *
 * <p>
 * The only constructor
 *
 * @param newSensorImpl The SensorImpl
 * @param newCurrentValue The starting value for the currentValue
 * @param newReadInterval The readInterval
 */
public AbstractSensor(SensorImpl newSensorImpl,
                int newCurrentValue, int newReadInterval) {
        setSensorImpl(newSensorImpl);
        sensorValue = newCurrentValue;
        setReadInterval(newReadInterval);
}

/**
 * getValue
 *
 * <p>
 * Returns the integer value of the sensor as last read
 */
public int getValue() {
        return sensorValue;
}

/**
 * getObjectValue
 *
 * <p>
 * Returns the object value of the sensor
 */
public Object getObjectValue() {
        return objectValue;
}

/**
 * setReadInterval
 */
```

```
public void setReadInterval(int newReadInterval) {
        readIntervalMillis = newReadInterval;
}

/**
 * getReadInterval
 */
public int getReadInterval() {
        return readIntervalMillis;
}

/**
 * addSensorFilter
 *
 * @param newFilter The newest sensorFilter to add
 */
public void addSensorFilter(SensorFilter newFilter) {
        if (filter == null) {
                // If this is the first filter, just set it
                filter = newFilter;
        } else {
                // If there is already a filter, add the filter to the chain
                filter.addFilter(filter);
        }
}

/**
 * removeSensorFilter
 *
 * @param oldFilter The filter to remove
 */
public void removeSensorFilter(SensorFilter oldFilter) {
        if (filter != null) {
                filter = filter.removeFilter(oldFilter);
        }
        else if (filter == oldFilter) {
                filter = null;
        }
}

/**
 * filterSensor
```

```
 *
 * <p>
 * Apply the chain of SensorFilter's to the Sensor
 */
public void filterSensor() {
        if (filter != null) {
                filter.filterReading(sensorValue);
        }
}

/**
 * addSensorListener
 */
public void addSensorListener(SensorListener listener) {
        if (!listeners.contains(listener))  {
                listeners.add(listener);
        }
}

/**
 * removeSensorListener
 */
public void removeSensorListener(SensorListener listener) {
        listeners.remove(listener);
}

/**
 * setValue
 *
 * <p>
 * Set the integer value, filter it, and let the Sensor know if
 * it's changed.
 */
public void setValue(int newValue) {
        int oldValue = sensorValue;
        sensorValue = newValue;
        filterSensor();
        setChanged(oldValue != newValue);
}

/**
 * setObjectValue
```

```
 *
 * <p>
 * Set the object vlaue of the Sensor
 */
public void setObjectValue(Object o)  {
          objectValue = o;
}

/**
 * fireSensorListeners()
 */
public void fireSensorListeners() {
          // This is simplistic: it should be done within its own thread
          SensorEvent event = new SensorEvent(this);
          Iterator iter = listeners.iterator();
          while (iter.hasNext()) {
                    SensorListener listener = (SensorListener) iter.next();
                    listener.sensorChanged(event);
          }
}

/**
 * shouldRead
 *
 * <p>
 * This is true if the sensor is ready to be read
 */
public boolean shouldRead(long time) {
          return time >= nextReadTime;
}

/**
 * isChanged
 *
 * @return true if the sensor has changed
 */
public boolean isChanged()  {
          return sensorChanged;
}

/**
 * setChanged
```

```
     *
     * @param b The new value
     */
    public void setChanged(boolean b) {
            sensorChanged = b;
    }

    /**
     * readSensor
     *
     * <p>
     * Handle the basics of reading the sensor
     */
    public void readSensor(long time) {
            // Set the times
            lastReadTime = time;
            nextReadTime = time + readIntervalMillis;

            // Read the hardware implementation of the sensor
            sensorImpl.readSensor();

            // Set the integer and object values
            setObjectValue(sensorImpl.getObjectValue());
            setValue(sensorImpl.getIntValue());

            // If the sensor is changed, fire the listeners
            if (isChanged()) {
                    fireSensorListeners();
                    setChanged(false);
            }
    }

    /**
     * setSensorImpl
     *
     * <p>
     * Set the SensorImpl
     *
     * @see ws.enerd.robots.Sensor#setSensorImpl(ws.enerd.robots.sen-
sor.SensorImpl)
     */
    public void setSensorImpl(SensorImpl impl) {
```

```
			sensorImpl = impl;
	}

	/**
	 * setActive
	 *
	 * <p>
	 * Change the activity of the sensor
	 *
	 * @see ws.enerd.robots.Sensor#setActive(boolean)
	 */
	public void setActive(boolean active) {
			this.active = active;
	}

	/**
	 * isActive
	 *
	 * <p>
	 * Returns the activity of the sensor
	 *
	 * @see ws.enerd.robots.Sensor#isActive()
	 */
	public boolean isActive() {
			return active;
	}
}
```

ws.enerd.robots.sensor.BooleanSensor.java

```
/*
 * Created on Jul 3, 2005
 *
 * The licence is the Attribution-ShareAlike 2.0 License as stated at:
 * http://creativecommons.org/licenses/by-sa/2.0/
 */
package ws.enerd.robots.sensor;

/**
 * BooleanSensor
 *
 * <p>
```

```
 * A Sensor that returns either a true or a false.
 *
 * <p>
 * Copyright 2005 D. Jay Newman
 *
 * <p>
 * The licence is the Attribution-ShareAlike 2.0 License as stated at:
 * http://creativecommons.org/licenses/by-sa/2.0/
 *
 * @author D. Jay Newman
 */
public class BooleanSensor extends AbstractSensor {

    public BooleanSensor(SensorImpl newSensorImpl, int newCurrentValue,
                         int newReadInterval) {
            super(newSensorImpl, newCurrentValue, newReadInterval);
    }

    /**
     * getBooleanValue
     *
     * <p>
     * Returns true if the sensor value is non-zero
     */
    public boolean getBooleanValue() {
            return sensorValue != 0;
    }
}
```

ws.enerd.robots.sensor.DistanceSensor.java

```
/*
 * Created on Jul 2, 2005
 *
 * The licence is the Attribution-ShareAlike 2.0 License as stated at:
 * http://creativecommons.org/licenses/by-sa/2.0/
 */
package ws.enerd.robots.sensor;

/**
 * DistanceSensor
 *
```

```
 * <p>
 * A class which deals with distance as a simple number and
 * allows the definition of distance ranges (close, medium, and
 * far).
 *
 * <p>
 * The values for close and far must be determined on the basis of
 * use rather than strictly by sensor. For example, a sensor designed
 * to detect dog toys that might block a wheel might have a different
 * set of ranges than a sensor designed to detect open doorways.
 *
 * <p>
 * If proportional is true then the distance is propotional to the
 * value of the sensor; otherwise the distance is inversely proportional.
 *
 * <p>
 * The Sharp IR Rangers are inversely proportional to the value
 *
 * <p>
 * Copyright 2005 D. Jay Newman
 *
 * <p>
 * The licence is the Attribution-ShareAlike 2.0 License as stated at:
 * http://creativecommons.org/licenses/by-sa/2.0/
 *
 * @author D. Jay Newman
 */
public class DistanceSensor extends AbstractSensor {
   protected int close = 0;
   protected int far = 0;

   // If proportional is true then the range is proportional to
   // the sensor value; if false then it is inversely proportional
   protected boolean proportional = true;

   public DistanceSensor(SensorImpl newSensorImpl, int
newCurrentValue,
                      int newReadInterval) {
         super(newSensorImpl, newCurrentValue, newReadInterval);
   }

   /**
```

```
     * DistanceSensor
     *
     * <p>
     * The new constructor that includes the range definitions
     *
     * @param newSensorImpl
     * @param newCurrentValue
     * @param newReadInterval
     * @param newCloseValue
     * @param newFarValue
     */
    public DistanceSensor(SensorImpl newSensorImpl, int
newCurrentValue,
                    int newReadInterval, int newCloseValue, int
newFarValue) {
            this(newSensorImpl, newCurrentValue, newReadInterval);
            setCloseRange(newCloseValue);
            setFarRange(newFarValue);
    }

    /**
     * DistanceSensor
     *
     * @param newSensorImpl
     * @param newCurrentValue
     * @param newReadInterval
     * @param newCloseValue
     * @param newFarValue
     * @param newProportional
     */
    public DistanceSensor(SensorImpl newSensorImpl, int
newCurrentValue,
                    int newReadInterval, int newCloseValue, int
newFarValue,
                    boolean newProportional) {
            this(newSensorImpl, newCurrentValue, newReadInterval,
newCloseValue,
                            newFarValue);
            setProportional(newProportional);
    }

    /**
```

```
 * setCloseRange
 *
 * <p>
 * Everything less (or greater than if propotional is false) is
 * considered "close".
 *
 * @param r
 */
public void setCloseRange(int r) {
	this.close = r;
}

/**
 * setFarRange
 *
 * <p>
 * Everything greater (or less than if proportional is false) is
 * considered "far".
 *
 * @param r
 */
public void setFarRange(int r) {
	this.far = r;
}

/**
 * setProportional
 *
 * @param b
 */
public void setProportional(boolean b) {
	this.proportional = b;
}

/**
 * isClose
 *
 * @return
 */
public boolean isClose() {
	return proportional ? (sensorValue <= close) : (sensorValue >=
close);
```

```
    }

    /**
     * isFar
     *
     * @return
     */
    public boolean isFar() {
            return proportional ? (sensorValue >= far) : (sensorValue <=
far);
    }

    /**
     * isMedium
     *
     * <p>
     * This is true if the value is between close and far
     *
     * @return
     */
    public boolean isMedium() {
            return (sensorValue > close) && (sensorValue < far);
    }
}
```

ws.enerd.robots.sensor.StringSensor.java

```
/*
 * Created on Aug 6, 2005
 *
 * The licence is the Attribution-ShareAlike 2.0 License as stated at:
 * http://creativecommons.org/licenses/by-sa/2.0/
 */
package ws.enerd.robots.sensor;

/**
 * StringSensor
 *
 * <p>
 * This is a sensor that is triggered by something that
 * sends a String. I have used this both for the Sphinx4 voice
 * recognition system and a JSP remote controller.
```

```
 *
 * <p>
 * Copyright 2005 D. Jay Newman
 *
 * <p>
 * The licence is the Attribution-ShareAlike 2.0 License as stated at:
 * http://creativecommons.org/licenses/by-sa/2.0/
 *
 * @author D. Jay Newman
 */
public class StringSensor extends AbstractSensor {

    /**
     * @param newSensorImpl
     * @param newReadInterval
     */
    public StringSensor(SensorImpl newSensorImpl, int newReadInterval) {
            super(newSensorImpl, 0, newReadInterval);
    }

    /**
     * setValue
     *
     * <p>
     * This is called by readSensor and it comes from
     * the SensorImpl
     */
    public void setValue(int newValue) {
            sensorValue = newValue;
    }

    /**
     * setChanged
     *
     * <p>
     * The reading is considered changed anytime the
     * new objectValue (o) is different from objectValue
     */
    public void setObjectValue(Object o)  {
            if (o != objectValue) {
                    objectValue = o;
                    setChanged(true);
```

```
                }
        }

        /**
         * getString
         *
         * <p>
         * As a side effect, getString also sets the objectValue to
         * null. This method is typically called from a SensorListener
         * after the value has been changed.
         *
         * @return A String containing the value of the sensor
         */
        public String getString() {
                String s = (String)objectValue;
                objectValue = null;
                return s;
        }
}
```

ws.enerd.robots.sensor.TPAData.java

```
/*
 * Created on Jul 13, 2005
 *
 * The licence is the Attribution-ShareAlike 2.0 License as stated at:
 * http://creativecommons.org/licenses/by-sa/2.0/
 */
package ws.enerd.robots.sensor;

/**
 * TPAData
 *
 * <p>
 * The data from reading the Devantech TPA
 *
 * <p>
 * Copyright 2005 D. Jay Newman
 *
 * <p>
 * The licence is the Attribution-ShareAlike 2.0 License as stated at:
 * http://creativecommons.org/licenses/by-sa/2.0/
```

```
 *
 * @author D. Jay Newman
 */
public class TPAData {
    // The ambient temperature
    private int ambient = 0;

    // The individual temperature pixels
    private int[] pixels = new int[8];

    /**
     * TPAData
     */
    public TPAData() {
            for (int i = 0; i < 8; i++) {
                    setPixel(i, 0);
            }
    }

    public void setAmbient(int i) {
            ambient = i;
    }

    public int getAmbient() {
            return ambient;
    }

    public void setPixel(int pixel, int value) {
            pixels[pixel] = value;
    }

    public int getPixel(int pixel) {
            return pixels[pixel];
    }
}
```

ws.enerd.robots.sensor.TPASensor.java

```
/*
 * Created on Jul 13, 2005
 *
 * The licence is the Attribution-ShareAlike 2.0 License as stated at:
```

```
 * http://creativecommons.org/licenses/by-sa/2.0/
 */
package ws.enerd.robots.sensor;

import ws.enerd.robots.mojo.MojoConnection;

/**
 * TPASensor
 *
 * <p>
 * A Sensor class that reads a Deventech TPA (Thermopile Array)
 *
 * <p>
 * This is still experimental
 *
 * <p>
 * Copyright 2005 D. Jay Newman
 *
 * <p>
 * The licence is the Attribution-ShareAlike 2.0 License as stated at:
 * http://creativecommons.org/licenses/by-sa/2.0/
 *
 * @author D. Jay Newman
 */
public class TPASensor extends AbstractSensor {

    protected MojoConnection connection = null;

    /**
     * @param newSensorImpl
     * @param newCurrentValue
     * @param newReadInterval
     * @param mojoConnection
     */
    public TPASensor(SensorImpl newSensorImpl, int newCurrentValue,
                        int newReadInterval, MojoConnection
mojoConnection) {
              super(newSensorImpl, newCurrentValue, newReadInterval);
              this.connection = mojoConnection;
    }
}
```

ws.enerd.robots.sensor.SensorFilter.java

```
/*
 * Created on Mar 26, 2005
 *
 * The licence is the Attribution-ShareAlike 2.0 License as stated at:
 * http://creativecommons.org/licenses/by-sa/2.0/
 */
package ws.enerd.robots.sensor;

/**
 * SensorFilter.java
 *
 * <p>
 * An interface to a generic sensor filter
 *
 * <p>
 * The licence is the Attribution-ShareAlike 2.0 License as stated at:
 * http://creativecommons.org/licenses/by-sa/2.0/
 *
 * @author D. Jay Newman
 */
public interface SensorFilter {
    // This method should be overridden
    public int doFilterReading(int newReading);

    // This handles recursion and then calls doFilterReading(int)
    public int filterReading(int newReading);

    // Allow filters to be chained; added filters are called first.
    public void addFilter(SensorFilter newFilter);

    public SensorFilter removeFilter(SensorFilter oldFilter);
}
```

ws.enerd.robots.sensor.AbstractSensorFilter.java

```
/*
 * Created on Mar 26, 2005
 *
 * The licence is the Attribution-ShareAlike 2.0 License as stated at:
 * http://creativecommons.org/licenses/by-sa/2.0/
```

```
 */
package ws.enerd.robots.sensor;

/**
 * AbstractSensorFilter
 *
 * <p>
 * An implementation of the SensorFilter interface; this implementation
 * handles the nitty-gritty of dealing with other SensorFilter's
 *
 * <p>
 * The licence is the Attribution-ShareAlike 2.0 License as stated at:
 * http://creativecommons.org/licenses/by-sa/2.0/
 *
 * @author D. Jay Newman
 */
public abstract class AbstractSensorFilter implements SensorFilter {
   protected SensorFilter nextFilter = null;

   /**
    * filterReading
    *
    * <p>
    * First check the next filter in the chain, then do the
    * filtering
    */
   public int filterReading(int newReading) {
            if (nextFilter != null)  {
                     newReading = filterReading(newReading);
            }

            return doFilterReading(newReading);
   }

   /**
    * doFilterReading
    *
    * <p>
    * This must be overridden because this is the main
    * reason for this class
    */
   public abstract int doFilterReading(int newReading);
```

```
	/**
	 * addFilter
	 *
	 * <p>
	 * This is easy: just recursively follow the list and add the
	 * filter to the end of the list.
	 */
	public void addFilter(SensorFilter newFilter) {
		if (nextFilter == null) {
			nextFilter = newFilter;
		}
		else {
			nextFilter.addFilter(newFilter);
		}
	}


	/**
	 * removeFilter
	 *
	 * <p>
	 * Find the filter to remove and, well, remove it. Then remake
	 * the chain.
	 */
	public SensorFilter removeFilter(SensorFilter oldFilter) {
		if (oldFilter == this) {
			return nextFilter;
		}
		else {
			if (nextFilter == null) {
				return null;
			}
			else {
				nextFilter = nextFilter.removeFilter(oldFilter);
			}
		}

		return this;
	}
}
```

ws.enerd.robots.sensor.AvergingArrayFilter.java

```
/*
 * Created on Mar 28, 2005
 *
 * The licence is the Attribution-ShareAlike 2.0 License as stated at:
 * http://creativecommons.org/licenses/by-sa/2.0/
 */
package ws.enerd.robots.sensor;

/**
 * AveragingArrayFilter
 *
 * <p>
 * A Filter that averages the current reading with previous
 * readings
 *
 * <p>
 * The licence is the Attribution-ShareAlike 2.0 License as stated at:
 * http://creativecommons.org/licenses/by-sa/2.0/
 *
 * @author D. Jay Newman
 */
public class AveragingArrayFilter extends AbstractSensorFilter {

    // The circular array of values
    private int[] values;
    private int len = 0;
    private int nextIndex = 0;

    public AveragingArrayFilter(short length) {
            values = new int[length];
            for (int i = 0; i < length; i++) {
                    values[i] = 0;
            }
    }

    /**
     * doFilterReading
     *
     * <p>
     * Add the value to the array and return the new average.
```

```
     *
     * @see ws.enerd.robots.sensor.SensorFilter#doFilterReading(int)
     */
    public int doFilterReading(int newReading) {
        values[nextIndex++] = newReading;
        if (nextIndex >= values.length) {
            nextIndex = 0;
        }

        int sum = 0;
        for (int i = 0; i < values.length; i++) {
            sum += values[i];
        }

        return sum / values.length;
    }

}
```

ws.enerd.robots.sensor.SensorImpl.java

```
/*
 * Created on May 28, 2005
 *
 * The licence is the Attribution-ShareAlike 2.0 License as stated at:
 * http://creativecommons.org/licenses/by-sa/2.0/
 */
package ws.enerd.robots.sensor;

/**
 * SensorImpl
 *
 * <p>
 * The interface between a sensor type and the hardware behind it.
 *
 * <p>
 * Copyright 2005 D. Jay Newman
 *
 * <p>
 * The licence is the Attribution-ShareAlike 2.0 License as stated at:
 * http://creativecommons.org/licenses/by-sa/2.0/
 *
```

```
 * @author D. Jay Newman
 */
public interface SensorImpl {
    public void readSensor();
    public int getIntValue();
    public Object getObjectValue();
}
```

ws.enerd.robots.sensor.AbstractSensorImpl.java

```
/*
 * Created on May 28, 2005
 *
 * The licence is the Attribution-ShareAlike 2.0 License as stated at:
 * http://creativecommons.org/licenses/by-sa/2.0/
 */
package ws.enerd.robots.sensor;

/**
 * AbstractSensorImpl
 *
 * <p>
 * An interface of SensorImpl that handles the boring stuff
 *
 * <p>
 * Copyright 2005 D. Jay Newman
 *
 * <p>
 * The licence is the Attribution-ShareAlike 2.0 License as stated at:
 * http://creativecommons.org/licenses/by-sa/2.0/
 *
 * @author D. Jay Newman
 */
public abstract class AbstractSensorImpl implements SensorImpl {
    protected int intValue = 0;
    protected Object objectValue = null;

    /**
     * readSensor
     *
     * <p>
     * This does the job of actually reading the sensor and must
```

```
	 * be overridden.
	 *
	 * @see ws.enerd.robots.sensor.SensorImpl#readSensor()
	 */
	abstract public void readSensor();

	/**
	 * getIntValue
	 *
	 * @see ws.enerd.robots.sensor.SensorImpl#getIntValue()
	 */
	public int getIntValue() {
		return intValue;
	}

	/**
	 * getObjectValue
	 *
	 * @see ws.enerd.robots.sensor.SensorImpl#getObjectValue()
	 */
	public Object getObjectValue() {
		return objectValue;
	}
}
```

ws.enerd.robots.sensor.FileSensorImpl.java

```
/*
 * Created on Jul 2, 2005
 *
 * The licence is the Attribution-ShareAlike 2.0 License as stated at:
 * http://creativecommons.org/licenses/by-sa/2.0/
 */
package ws.enerd.robots.sensor;

import java.io.BufferedReader;
import java.io.File;
import java.io.FileNotFoundException;
import java.io.FileReader;
import java.io.IOException;

/**
```

```
 * FileSensorImpl
 *
 * <p>
 * A SensorImpl that reads a file to get the value
 *
 * <p>
 * Copyright 2005 D. Jay Newman
 *
 * <p>
 * The licence is the Attribution-ShareAlike 2.0 License as stated at:
 * http://creativecommons.org/licenses/by-sa/2.0/
 *
 * @author D. Jay Newman
 */
public class FileSensorImpl extends AbstractSensorImpl {
    File sensorFile = null;

    public FileSensorImpl(String path)  {
            sensorFile = new File(path);
    }

    /**
     * readSensor
     *
     * <p>
     * Whenever this is called, the file is read and the value is
     * assumed to be an int
     *
     * @see ws.enerd.robots.sensor.SensorImpl#readSensor()
     */
    public void readSensor() {
            String s = null;
            try {
                    // Open and read the file from the File object
                    BufferedReader reader = new BufferedReader(new
FileReader(sensorFile));
                    s = reader.readLine();
                    reader.close();
                    intValue = Integer.parseInt(s);
            } catch (FileNotFoundException e) {
                    e.printStackTrace();
            } catch (IOException e) {
```

```
                    e.printStackTrace();
               }
          }
     }
```

ws.enerd.robots.sensor.MojoSensorRingImpl.java

```
/*
 * Created on Jun 2, 2005
 *
 * The licence is the Attribution-ShareAlike 2.0 License as stated at:
 * http://creativecommons.org/licenses/by-sa/2.0/
 */
package ws.enerd.robots.sensor;

import ws.enerd.robots.mojo.MojoSensorCircle;

/**
 * MojoSensorRingImpl
 *
 * <p>
 * This is a SensorImpl that is specific to any robot that has
 * a ring of sonar units controlled by a MojoBus device with my
 * specific set of commands. The MojoSensorCircle does the
 * actual interfacing with the sonar units.
 *
 * <p>
 * Copyright 2005 D. Jay Newman
 *
 * <p>
 * The licence is the Attribution-ShareAlike 2.0 License as stated at:
 * http://creativecommons.org/licenses/by-sa/2.0/
 *
 * @author D. Jay Newman
 */
public class MojoSensorRingImpl extends AbstractSensorImpl {
   private MojoSensorCircle circle = null;
   private int index = 0;


   public MojoSensorRingImpl(MojoSensorCircle msc, int i)  {
          circle = msc;
```

```
            index = i;

            objectValue = null;
    }

    /**
     * readSensor
     *
     * @see ws.enerd.robots.sensor.SensorImpl#readSensor()
     */
    public void readSensor() {
            intValue = circle.getSensorData(index);
    }
}
```

ws.enerd.robots.sensor.SocketSensorImpl.java

```
/*
 * Created on Aug 3, 2005
 *
 * The licence is the Attribution-ShareAlike 2.0 License as stated at:
 * http://creativecommons.org/licenses/by-sa/2.0/
 */
package ws.enerd.robots.sensor;

import java.io.BufferedReader;
import java.io.IOException;
import java.io.InputStreamReader;
import java.net.Socket;
import java.net.UnknownHostException;

/**
 * SocketSensorImpl
 *
 * <p>
 * A SensorImpl that opens a socket to read a string
 *
 * <p>
 * Copyright 2005 D. Jay Newman
 *
 * <p>
 * The licence is the Attribution-ShareAlike 2.0 License as stated at:
```

```
 * http://creativecommons.org/licenses/by-sa/2.0/
 *
 * @author D. Jay Newman
 */
public class SocketSensorImpl extends AbstractSensorImpl {
   protected String host = null;
   protected int port = 0;
   protected Socket socket = null;
   protected BufferedReader reader = null;

   public SocketSensorImpl(String h, int p) {
            this.host = h;
            this.port = p;

            openSocket();
   }

   /**
    * openSocket
    *
    * <p>
    * Open the socket to the server if it isn't already open
    *
    */
   public void openSocket() {
            if ((socket != null) && socket.isConnected()) {
                     return;
            }

            // Now open the socket
            try {
                     socket = new Socket(host, port);
                     reader = new BufferedReader(new
InputStreamReader(socket
                                          .getInputStream()));

            } catch (UnknownHostException e) {
                     e.printStackTrace();
            } catch (IOException e) {
                     e.printStackTrace();
            }
   }
```

```
    /**
     * readSensor
     *
     * @see ws.enerd.robots.sensor.SensorImpl#readSensor()
     */
    public void readSensor() {
            String s = null;

            openSocket();

            try {
                    if (reader.ready()) {
                            s = reader.readLine();
                    }
                    if (s != null) {
                            s.trim();
                    }

                    objectValue = s;

                    intValue = ((s == null) || (s.length() == 0)) ? 0 : 1;
            } catch (IOException e) {
                    e.printStackTrace();
            }
    }

}
```

ws.enerd.robots.sensor.SocketServerSensorImpl.java

```
/*
 * Created on Jul 28, 2005
 *
 * The licence is the Attribution-ShareAlike 2.0 License as stated at:
 * http://creativecommons.org/licenses/by-sa/2.0/
 */
package ws.enerd.robots.sensor;

import java.io.BufferedReader;
import java.io.IOException;
import java.io.InputStreamReader;
```

```
import java.net.ServerSocket;
import java.net.Socket;
import java.util.Vector;

/**
 * SocketSensorImpl
 *
 * <p>
 * A SensorImpl that acts as a server for some some other application.
 *
 * <p>
 * This creates a trivial server that allows another application to
 * open a socket and send data
 *
 * <p>
 * Copyright 2005 D. Jay Newman
 *
 * <p>
 * The licence is the Attribution-ShareAlike 2.0 License as stated at:
 * http://creativecommons.org/licenses/by-sa/2.0/
 *
 * @author D. Jay Newman
 */
public class SocketServerSensorImpl extends AbstractSensorImpl {
   class ServerThread extends Thread {
            ServerSocket socket = null;
            Vector lines = new Vector();

            public ServerThread(int p) {
                     try {
                              socket = new ServerSocket(p);
                     } catch (IOException ex) {
                              ex.printStackTrace();
                     }
            }

            public void run() {
                     while (true) {
                              try {
                                       Socket client = socket.accept();
                                       BufferedReader reader = new
BufferedReader(
```

```
                                                    new
InputStreamReader(client.getInputStream()));
                                    while (!client.isClosed()) {
                                          String s =
reader.readLine();
                                    }
                              } catch (IOException ex) {
                                    ex.printStackTrace();
                              }
                              Thread.yield();
                        }
                  }

                  public String getFirstLine() {
                        String s = null;

                        synchronized (lines) {
                              if (lines.size() > 0) {
                                    s = (String) lines.get(0);
                                    lines.removeElementAt(0);
                              }
                        }

                        return s;
                  }
            }

            protected ServerThread thread = null;

            /**
             * SocketServerSensorImpl
             *
             * @param p port number
             */
            public SocketServerSensorImpl(int p) {
                  thread = new ServerThread(p);
            }

            /**
             * readSensor
             *
             * <p>
```

```
     * Get the current string that is being read for
     *
     * @see ws.enerd.robots.sensor.SensorImpl#readSensor()
     */
    public void readSensor() {
            String s = thread.getFirstLine();
            objectValue = s;
            if (s == null) {
                    intValue = 0;
            } else {
                    intValue = 1;
            }
    }
}
```

ws.enerd.robots.sensor.SensorEvent.java

```
/*
 * Created on Mar 23, 2005
 *
 * The licence is the Attribution-ShareAlike 2.0 License as stated at:
 * http://creativecommons.org/licenses/by-sa/2.0/
 */
package ws.enerd.robots.sensor;

import java.util.EventObject;
import ws.enerd.robots.Sensor;

/**
 * SensorEvent
 *
 * <p>
 * This is used to send data to a SensorListener
 *
 * @author D. Jay Newman
 *
 *<p>
 * The licence is the Attribution-ShareAlike 2.0 License as stated at:
 * http://creativecommons.org/licenses/by-sa/2.0/
 *
 */
public class SensorEvent extends EventObject {
```

```
    public SensorEvent(Sensor source) {
        super(source);
    }
}
```

ws.enerd.robots.sensor.SensorListener.java

```
/*
 * Created on Mar 23, 2005
 *
 * The licence is the Attribution-ShareAlike 2.0 License as stated at:
 * http://creativecommons.org/licenses/by-sa/2.0/
 */
package ws.enerd.robots.sensor;

/**
 * SensorListener
 *
 * <p>
 * This is a very basic listener for SensorEvent's
 *
 * @author D. Jay Newman
 *
 * <p>
 * The licence is the Attribution-ShareAlike 2.0 License as stated at:
 * http://creativecommons.org/licenses/by-sa/2.0/
 *
 */
public interface SensorListener {
    public void sensorChanged(SensorEvent e);
}
```

Behavioral Programming

Behavioral programming is an easy way of programming a robot to react in real time to its surroundings.

The major weakness in my framework is that all my Arbitrators only allow for one active behavior at a time. So far this hasn't caused me any problems.

ws.enerd.robots.Behavior.java

```
/*
 * Created on Mar 29, 2005
 *
 * The licence is the Attribution-ShareAlike 2.0 License as stated at:
 * http://creativecommons.org/licenses/by-sa/2.0/
 */
package ws.enerd.robots.behavior;

import ws.enerd.robots.sensor.SensorListener;

/**
 * Behavior
 *
 * <p>
 * The interface for a Behavior for Behavioral Programming
 *
 * <p>
 * Copyright 2005 D. Jay Newman
 *
 * <p>
 * The licence is the Attribution-ShareAlike 2.0 License as stated at:
 * http://creativecommons.org/licenses/by-sa/2.0/
 *
 * @author D. Jay Newman
 */
public interface Behavior extends SensorListener {
   // Called once when the behavior is started
   public void start();

   // Called repeatedly while behavior is running; this
   // should not take a long time to complete
   public void run();

   // Called once when the behavior is stopped
   public void stop();

   // Called by the arbitration when the behavior should stop
   public void shouldStop();

   // Returns true when the behavior is stopped
```

```
    public boolean isStopped();

    // Returns when this behavior is fully stopped
    public void waitForStop();

    // Forces the behavior to wait before stopping
    public void wait(int millis);

    // Returns an integer corresponding with how much this
    // behavior wants to run at this time
    public int getActivation();

    public void setActivation(int i);

    public void addBehaviorListener(BehaviorListener listener);

    public void removeBehaviorListener(BehaviorListener listener);
}
```

ws.enerd.robots.AbstractBehavior.java

```
/*
 * Created on Mar 29, 2005
 *
 * The licence is the Attribution-ShareAlike 2.0 License as stated at:
 * http://creativecommons.org/licenses/by-sa/2.0/
 */
package ws.enerd.robots.behavior;

import java.util.ArrayList;
import java.util.Iterator;

import ws.enerd.robots.sensor.SensorEvent;

/**
 * AbstractBehavior
 *
 * <p>
 * Copyright 2005 D. Jay Newman
 *
 * <p>
 * The licence is the Attribution-ShareAlike 2.0 License as stated at:
```

```
 * http://creativecommons.org/licenses/by-sa/2.0/
 *
 * @author D. Jay Newman
 */
public abstract class AbstractBehavior implements Behavior {
    protected boolean running = true;
    protected boolean stopped = false;

    protected long wakeupTime = 0;

    protected int activation = 0;

    protected ArrayList behaviorListeners = new ArrayList();

    public AbstractBehavior() {
    }

    /**
     * start()
     *
     * <p>
     * Called to start a behavior running; super.start() should be called
     * if this method is overridden.
     *
     * @see ws.enerd.robots.behavior.Behavior#start()
     */
    public void start() {
                running = true;
                stopped = false;
                wakeupTime = 0;
    }

    /**
     * run()
     *
     * <p>
     * Called when the behavior should be running. This should be a fairly
short
     * method so that other Behavior's will get a chance to run if necessary
or
     * else you should sprinkle Thread.yield()'s througout.
     *
```

```
     * <p>
     * Either a subclass should call super.run() or handle the stopping code
     * itself.
     *
     * @see ws.enerd.robots.behavior.Behavior#run()
     */
    public void run() {
            if (running == false) {
                    stopped = true;
            }
    }

    /**
     * stop()
     *
     * <p>
     * Called when a behavior is stopped for whatever reason (usually
because a
     * new behavior will be started).
     *
     * @see ws.enerd.robots.behavior.Behavior#stop()
     */
    public void stop() {
    }

    /**
     * shouldStop
     *
     * <p>
     * This tells the Behavior that it should stop; this doesn't return
     * until the Behavior is stopped
     */
    public void shouldStop() {
            // Allow waiting to occur without interruption
            while (wakeupTime > System.currentTimeMillis()) {
                    try {
                            Thread.yield();
                    } catch (Exception ex) {
                    }
            }
            wakeupTime = 0;
```

```
		running = false;
	}

	public boolean isStopped() {
		return stopped;
	}

	public void waitForStop() {
		shouldStop();
		while (!isStopped()) {
			Thread.yield();
		}
	}

	/**
	 * wait
	 *
	 * <p>
	 * This tells the Behavior to wait for a certain amount of
	 * time before other Behavior's can be choosen.
	 */
	public void wait(int millis) {
		wakeupTime = System.currentTimeMillis() + millis;
	}

	/**
	 * getActivation
	 *
	 * <p>
	 * Currently I only use 0 or 1 here, but this could change
	 *
	 * @return The current activation level
	 */
	public int getActivation() {
		return activation;
	}

	/**
	 * setActivation
	 *
	 * <p>
	 * If the activation level changes, fire the listeners off the port bow
```

```
     */
    public void setActivation(int i) {
            int old_activation = activation;
            activation = i;
            if (old_activation != activation) {
                    fireBehaviorListeners();
            }
    }

    /**
     * sensorChanged
     *
     * <p>
     * This should be overriden if this Behavior cares about SensorEvent's
     *
     * @see
ws.enerd.robots.sensor.SensorListener#sensorChanged(SensorEvent)
     */
    public void sensorChanged(SensorEvent e) {
    }

    public void addBehaviorListener(BehaviorListener listener) {
            if (!behaviorListeners.contains(listener)) {
                    behaviorListeners.add(listener);
            }
    }

    public void removeBehaviorListener(BehaviorListener listener) {
            behaviorListeners.remove(listener);
    }

    protected void fireBehaviorListeners() {
            // This is simplistic: it should be done within its own thread
            Iterator iter = behaviorListeners.iterator();
            while (iter.hasNext()) {
                    BehaviorListener listener = (BehaviorListener)
iter.next();
                    listener.activityChanged(this);
            }
    }
}
```

ws.enerd.robots.NullBehavior.java

```
/*
 * Created on Jun 19, 2005
 *
 * The licence is the Attribution-ShareAlike 2.0 License as stated at:
 * http://creativecommons.org/licenses/by-sa/2.0/
 */
package ws.enerd.robots.behavior;

/**
 * NullBehavior
 *
 * <p>
 * A NullBehavior object is merely a behavior that does nothing
 * but the activation value is always 1. This can be used for the
 * final behavior of a SimpleSubsumptionArbitrator.
 *
 * <p>
 * Copyright 2005 D. Jay Newman
 *
 * <p>
 * The licence is the Attribution-ShareAlike 2.0 License as stated at:
 * http://creativecommons.org/licenses/by-sa/2.0/
 *
 * @author D. Jay Newman
 */
public class NullBehavior extends AbstractBehavior {

    /**
     *
     */
    public NullBehavior() {
            setActivation(1);
    }
}
```

ws.enerd.robots.BehaviorListener.java

```
/*
 * Created on Jun 18, 2005
 *
```

```
 * The licence is the Attribution-ShareAlike 2.0 License as stated at:
 * http://creativecommons.org/licenses/by-sa/2.0/
 */
package ws.enerd.robots.behavior;

/**
 * BehaviorListener
 *
 * <p>
 * Copyright 2005 D. Jay Newman
 *
 * <p>
 * The licence is the Attribution-ShareAlike 2.0 License as stated at:
 * http://creativecommons.org/licenses/by-sa/2.0/
 *
 * @author D. Jay Newman
 */
public interface BehaviorListener {
    public void activityChanged(Behavior b);
}
```

ws.enerd.robots.Arbitrator.java

```
/*
 * Created on Mar 29, 2005
 *
 * The licence is the Attribution-ShareAlike 2.0 License as stated at:
 * http://creativecommons.org/licenses/by-sa/2.0/
 */
package ws.enerd.robots.behavior;

/**
 * Arbitrator
 *
 * <p>
 * The interface for a generic Arbitrator used in behavioral programming
 *
 * <p>
 * Copyright 2005 D. Jay Newman
 *
 * <p>
 * The licence is the Attribution-ShareAlike 2.0 License as stated at:
```

```
 * http://creativecommons.org/licenses/by-sa/2.0/
 *
 * @author D. Jay Newman
 */
public interface Arbitrator extends BehaviorListener {
    public void addBehavior(Behavior newBehavior);

    public void removeBehavior(Behavior newBehavior);

    public void startArbitrator();

    public void stopBehavior();

    public void startBehavior(Behavior newBehavior);

    public void shouldStop();

    public boolean isStopped();

    public void waitForStop();

    public void chooseBehavior();

    public void chooseBehavior(Behavior behavior);

    /**
     * @see
ws.enerd.robots.sensor.BehaviorListener.activityChanged(ws.enerd.robots.b
ehavior.Behavior)
     */
    public abstract void activityChanged(Behavior b);

}
```

ws.enerd.robots.AbstractArbitrator.java

```
/*
 * Created on Mar 29, 2005
 *
 * The licence is the Attribution-ShareAlike 2.0 License as stated at:
 * http://creativecommons.org/licenses/by-sa/2.0/
 */
```

```
package ws.enerd.robots.behavior;

import java.util.ArrayList;
import ws.enerd.util.StoppableThread;

/**
 * Arbitrator
 *
 * <p>
 * This class defines a generic Arbitrator that can be subclassed for
 * different versions of arbitration.
 *
 * <p>
 * The only method that absolutely has to be overridden is
chooseBehavior().
 *
 * <p>
 * Copyright 2005 D. Jay Newman
 *
 * <p>
 * The licence is the Attribution-ShareAlike 2.0 License as stated at:
 * http://creativecommons.org/licenses/by-sa/2.0/
 *
 * @author D. Jay Newman
 */
public abstract class AbstractArbitrator implements Arbitrator,
        BehaviorListener {

    /**
     * ArbitratorThread
     *
     * <p>
     * This is a thread that stops and starts Behavior instances
     *
     * @author D. Jay Newman
     */
    private class ArbitratorThread extends StoppableThread {
        private Behavior behavior = null;

        // doRun(), doRun(), dododoRunRun()...
        public void doRun() {
            if (behavior != null) {
```

```
                    behavior.run();
                }
            }

            public void startBehavior(Behavior b) {
                if (b == behavior) {
                    return;
                }

                // Stop the currently running behavior
                stopBehavior();

                // Start the new behavior
                behavior = b;
                if (behavior != null) {
                    behavior.start();
                }
            }

            public void stopBehavior() {
                if (behavior != null) {
                    behavior.waitForStop();
                    behavior.stop();
                }
            }

            public boolean isBehaviorStopped() {
                return behavior.isStopped();
            }
        }

        // Instance variables

        // stopping is true while the current behaivor is still stopping
        protected boolean stopping = false;

        // A list of possible Behavior's
        protected ArrayList behaviors = new ArrayList();

        // The currently active Behavior
        protected Behavior activeBehavior = null;
```

```
// The ArbitratorThread for this Arbitrator
protected ArbitratorThread thread = null;

/**
 * The contructor: this provides the ArbitratorThread which
 * in the method used to change behaviors
 *
 */
public AbstractArbitrator() {
        thread = new ArbitratorThread();
}

/**
 * addBehavior
 *
 * <p>
 * Handle the scut-work of adding a Behavior, including assigning
 * the BehaviorListener
 *
 * @param behavior The Behavior to add
 */
public synchronized void addBehavior(Behavior behavior) {
        if (!behaviors.contains(behavior)) {
                behaviors.add(behavior);
                behavior.addBehaviorListener(this);
        }
}

/**
 * removeBehaivor
 *
 * <p>
 * Remove a Behavior safely, including removing the listener
 */
public synchronized void removeBehavior(Behavior behavior) {
        if (behavior == activeBehavior) {
                thread.stopBehavior();
        }
        behavior.removeBehaviorListener(this);
        behaviors.remove(behavior);
}
```

```
/**
 * stopBehavior
 *
 * <p>
 * Stop the currently running behavior
 */
public void stopBehavior() {
        thread.stopBehavior();
}

/**
 * startBehavior
 *
 * <p>
 * Start a new behavior.
 * <ul>
 * <li>Do nothing if the desired Behavior is already running</li>
 * <li>Otherwise set activeBehavior and get the BehaviorThread to
 * safely start the new Behavior</li>
 * </ul>
 *
 * @param b The Behavior to start
 */
public void startBehavior(Behavior b) {
        if (activeBehavior == b)
                return;

        activeBehavior = b;
        if (activeBehavior != null) {
                thread.startBehavior(activeBehavior);
        }
}

/**
 * startArbitrator
 *
 * <p>
 * Start the arbitrator properly with no active Behavior
 */
public void startArbitrator() {
        thread.start();
        chooseBehavior(null);
```

```
}

/**
 * shouldStop
 *
 * <p>
 * Tell the arbitrator to stop the currently running Behavior. Currently
 * this waits for the Behavior to stop before returning.
 */
public void shouldStop() {
        if (activeBehavior != null) {
                thread.stopBehavior();
        }
}

/**
 * isStopped
 *
 * @return true if there is not behavior running
 */
public boolean isStopped() {
        return (activeBehavior == null) || thread.isBehaviorStopped();
}

/**
 * waitForStop
 *
 * <p>
 * This is a way of stopping the currently running Behavior with
 * a guarentee that the Behavior is stopped before returning
 */
public void waitForStop() {
        shouldStop();
        while (!isStopped()) {
                Thread.yield();
        }
}

/**
 * chooseBehavior
 *
 * <p>
```

```
     * Set the currently running Behavior to null
     */
    public void chooseBehavior() {
            chooseBehavior(null);
    }

    /**
     * chooseBehavior
     *
     * <p>
     * "PikaBehavior, I choose you!"
     *
     * <p>
     * Seriously, this is the main method used to start a new
     * Behavior.
     *
     * <p>
     * As this is abstract, it <em>must</em> be overridden.
     */
    public abstract void chooseBehavior(Behavior b);

    /**
     * activityChanged
     *
     * <p>
     * This is the method of a SensorListener
     *
     * @see
ws.enerd.robots.sensor.BehaviorListener.activityChanged(ws.enerd.robots.b
ehavior.Behavior)
     */
    public void activityChanged(Behavior b) {
            chooseBehavior();
    }
}
```

ws.enerd.robots.ControlArbitrator.java

```
/*
 * Created on Jul 21, 2005
 *
 * The licence is the Attribution-ShareAlike 2.0 License as stated at:
```

```
 * http://creativecommons.org/licenses/by-sa/2.0/
 */
package ws.enerd.robots.behavior;

/**
 * RCBehavior
 *
 * <p>
 * An arbitrator to allow testing of a robot by remote control.
 * Basically whenever a Behavior requests control it will replace the
 * active behavior. If the behavior that sends the event
 * has an activation of 0, the behavior will be stopped.
 *
 * <p>
 * Copyright 2005 D. Jay Newman
 *
 * <p>
 * The licence is the Attribution-ShareAlike 2.0 License as stated at:
 * http://creativecommons.org/licenses/by-sa/2.0/
 *
 * @author D. Jay Newman
 */
public class ControlArbitrator extends AbstractArbitrator {

    /**
     * An empty constructor that just goes to the constructor for
     * AbstractBehavior
     */
    public ControlArbitrator() {
    }

    /**
     * chooseBehavior
     *
     * <p>
     * This is the major method that needs to be overridden. When a new
     * Behavior is choosen, the old one stops.
     *
     * @see
ws.enerd.robots.behavior.Arbitrator#chooseBehavior(ws.enerd.robots.behav
ior.Behavior)
     */
```

```
    public void chooseBehavior(Behavior b) {
        if (b == null) {
            stopBehavior();
            return;
        }

        if (b.getActivation() > 0) {
            startBehavior(b);
        }
        else {
            stopBehavior();
        }
    }
}
```

ws.enerd.robots.SimpleSubsumptionArbitrator.java

```
/*
 * Created on Jun 11, 2005
 *
 * The licence is the Attribution-ShareAlike 2.0 License as stated at:
 * http://creativecommons.org/licenses/by-sa/2.0/
 */
package ws.enerd.robots.behavior;

import java.util.*;

/**
 * SimpleSubsumptionArbitrator
 *
 * <p>
 * An arbitrator that implements a very simplified version of
 * Rodney Brooks' Subsumption Architechture.
 *
 * <p>
 * Copyright 2005 D. Jay Newman
 *
 * <p>
 * The licence is the Attribution-ShareAlike 2.0 License as stated at:
 * http://creativecommons.org/licenses/by-sa/2.0/
 *
 * @author D. Jay Newman
```

```
 */
public class SimpleSubsumptionArbitrator extends AbstractArbitrator imple-
ments
        Arbitrator {

    /**
     *
     */
    public SimpleSubsumptionArbitrator() {
        super();
    }

    /**
     * chooseBehavior
     *
     * <p>
     * Ignore the behavior requresting the change.
     *
     * @see ws.enerd.robots.behavior.AbstractArbitrator#chooseBehavior()
     */
    public void chooseBehavior(Behavior behavior) {
        // When a sensor changes, I have to reevaluate the proper
behavior to
        // activate; to make this really simple, I'm not even going to
check
        // which sensor has changed, but to figure out the new behavior
from
        // the current values of all the behaviors. The winner is the first
        // behavior that wants to take control.

        Iterator iter = behaviors.iterator();
        while (iter.hasNext()) {
            Behavior b = (Behavior)iter.next();
            if (b.getActivation() > 0) {
                startBehavior(b);
                break;
            }
        }
    }
}
```

Imaging

This package just puts a front end on another imaging package. I will be working on porting MAVIS to Linux, and will eventually use this as my imaging base.

ws.enerd.robots.image.ImageMinder.java

```
/*
 * Created on Jun 27, 2005
 *
 * The licence is the Attribution-ShareAlike 2.0 License as stated at:
 * http://creativecommons.org/licenses/by-sa/2.0/
 */
package ws.enerd.robots.image;

import java.awt.Graphics2D;
import java.awt.Image;
import java.awt.image.BufferedImage;
import java.io.FileOutputStream;
import java.io.IOException;
import java.net.URL;

import javax.swing.ImageIcon;

import org.generation5.vision.EqualizeFilter;
import org.generation5.vision.Filter;
import org.generation5.vision.GaussianFilter;
import org.generation5.vision.GreyscaleFilter;
import org.generation5.vision.SobelEdgeDetectorFilter;

import com.sun.image.codec.jpeg.JPEGCodec;
import com.sun.image.codec.jpeg.JPEGImageEncoder;

/**
 * ImageMinder
 *
 * <p>
 * I ran out of ideas for a name!
 *
 * <p>
 * This is just an abstraction layer between the image processing
```

```
 * operations and other applications. In the future I would like
 * to use the MAVIS vision system (http://www.leafproject.org/)
 *
 * <p>
 * Copyright 2005 D. Jay Newman
 *
 * <p>
 * The licence is the Attribution-ShareAlike 2.0 License as stated at:
 * http://creativecommons.org/licenses/by-sa/2.0/
 *
 * @author D. Jay Newman
 */
public class ImageMinder {

    private static final Filter gausianFilter = new GaussianFilter();

    private static final Filter equalizeFilter = new EqualizeFilter();

    private static final Filter sobelFilter = new SobelEdgeDetectorFilter();

    /**
     * getImage
     */
    public static Image getImage(URL url) {
            ImageIcon icon = new ImageIcon(url);
            return icon.getImage();
    }

    public static BufferedImage getBufferedImage(Image image) {
            BufferedImage b = new BufferedImage(image.getWidth(null),
                            image.getHeight(null),
BufferedImage.TYPE_INT_RGB);
            Graphics2D g = b.createGraphics();
            g.drawImage(image, 0, 0, null);

            return b;
    }

    public static BufferedImage getBufferedImage(URL url) {
            return getBufferedImage(getImage(url));
    }
```

```
    public static void writeJPEGFile(BufferedImage image, String path)
                    throws IOException {
            FileOutputStream out = new FileOutputStream(path);
            JPEGImageEncoder encoder =
JPEGCodec.createJPEGEncoder(out);
            encoder.encode(image);
            out.flush();
            out.close();
    }

    public static BufferedImage getGrayscaleImage(BufferedImage color)  {
            BufferedImage gray =
                    new BufferedImage(color.getWidth(), color.getHeight(),

BufferedImage.TYPE_BYTE_GRAY);
            GreyscaleFilter.toGrey(color, gray);
            return gray;
    }

    public static BufferedImage gausain(BufferedImage image)  {
            return gausianFilter.filter(image, null);
    }

    public static BufferedImage equalize(BufferedImage image) {
            return equalizeFilter.filter(image);
    }

    public static BufferedImage sobel(BufferedImage image) {
            return sobelFilter.filter(image);
    }

}
```

Speech

This is a simple interface to the Festival Text-to-Speech server.

```
/*
 * Created on Jul 30, 2005
 *
```

```
 * The licence is the Attribution-ShareAlike 2.0 License as stated at:
 * http://creativecommons.org/licenses/by-sa/2.0/
 */
package ws.enerd.robots.speech;

import java.io.BufferedReader;
import java.io.IOException;
import java.io.InputStream;
import java.io.InputStreamReader;
import java.io.OutputStreamWriter;
import java.net.Socket;

/**
 * SpeakThread
 *
 * <p>
 * This class sends data to a Festival speech server.
 *
 * <p>
 * Copyright 2005 D. Jay Newman
 *
 * <p>
 * The licence is the Attribution-ShareAlike 2.0 License as stated at:
 * http://creativecommons.org/licenses/by-sa/2.0/
 *
 * @author D. Jay Newman
 */
public class SpeakThread extends Thread {

    private String host = "localhost";
    private int port = 1314;
    private Socket socket = null;
    private OutputStreamWriter fest_in = null;
    private InputStream fest_out = null;

    private String text = null;

    /**
     *
     */
    public SpeakThread(String host, int port) {
            this.host = host;
```

```
			this.port = port;

			openSocket();
	}

	/**
	 * openSocket
	 *
	 * <p>
	 * Open the socket to the Festival server
	 */
	public void openSocket() {
			try {
					// Create the network socket
					socket = new Socket(host, port);

					// Create the writer to send data to festival
					fest_in = new
OutputStreamWriter(socket.getOutputStream());

					// Get the stream to get (and ignore) festival output
					fest_out = socket.getInputStream();
			} catch (Exception ex) {
					ex.printStackTrace();
			}
	}

	/**
	 * speak
	 *
	 * <p>
	 * Set the text to speak
	 * @param s
	 */
	public void speak(String s) {
			text = s;
	}

	/**
	 * run
	 *
	 * <p>
```

```
     * This is the core of the matter. A socket is opened and the
     * proper "SayText" command is sent to Festival. Then all the
     * output from Festival is read and ignored.
     */
    public void run() {
            while (true) {
                    try {
                            if (socket.isClosed()) {
                                    openSocket();
                            }

                            if (text != null) {
                                    fest_in.write("(SayText \"" + text +
"\")");

                                    fest_in.flush();
                                    text = null;
                            }

                            while (fest_out.available() > 0) {
                                    fest_out.read();
                            }
                    } catch (IOException ex) {
                            ex.printStackTrace();
                    }

                    Thread.yield();
            }
    }

    /*
     * This is here for testing purposes only
     */
    public static void main(String[] args)  {
            SpeakThread thread = new SpeakThread("groucho.home.jay",
1314);
            thread.start();

            BufferedReader console = new BufferedReader(new
InputStreamReader(System.in));

            String s = "";
            while (true)  {
```

```
                try {
                    s = console.readLine();
                } catch (IOException e) {
                    e.printStackTrace();
                }

                thread.speak(s);

                if (s.equals("exit")) {
                    System.exit(0);
                }

                Thread.yield();
            }
        }
    }
```

Mapping

These are the basic building blocks of an Occupancy Map.

ws.enerd.robots.mapping.OccupancyMap

```
/*
 * Created on Aug 7, 2005
 *
 * The licence is the Attribution-ShareAlike 2.0 License as stated at:
 * http://creativecommons.org/licenses/by-sa/2.0/
 */
package ws.enerd.robots.mapping;

import java.awt.Dimension;
import java.awt.geom.*;

/**
 * OccupancyMap
 *
 * <p>
 * A fairly basic Occupancy Map.
 *
 * <p>
 * Copyright 2005 D. Jay Newman
```

```
 *
 * <p>
 * The licence is the Attribution-ShareAlike 2.0 License as stated at:
 * http://creativecommons.org/licenses/by-sa/2.0/
 *
 * @author D. Jay Newman
 */
public class OccupancyMap {

    protected MapCell[][] map = null;
    protected Point2D cellSize = null;

    public OccupancyMap() {
    }

    public OccupancyMap(int x, int y) {
            this();
            map = new MapCell[x][y];
    }

    public void setMapAt(int x, int y, MapCell newCell) {
            map[x][y] = newCell;
    }

    public MapCell getCellAt(int x, int y) {
            if (map != null)  {
                    return map[x][y];
            }
            return null;
    }

    public Dimension getDimension() {
            Dimension d = null;
            if (map == null) {
                    d = new Dimension(0, 0);
            }
            else {
                    d = new Dimension(map.length, map[0].length);
            }

            return d;
    }
```

```
    public Point2D getCellSize() {
            return cellSize;
    }

    public void setCellSize(double x, double y) {
            cellSize = new Point2D.Double(x, y);
    }
}
```

ws.enerd.robots.mapping.MapCell

```
/*
 * Created on Aug 29, 2005
 *
 * The licence is the Attribution-ShareAlike 2.0 License as stated at:
 * http://creativecommons.org/licenses/by-sa/2.0/
 */
package ws.enerd.robots.mapping;

/**
 * MapCell
 *
 * <p>
 * A single cell of an OccupancyMap.
 *
 * <p>
 * Copyright 2005 D. Jay Newman
 *
 * <p>
 * The licence is the Attribution-ShareAlike 2.0 License as stated at:
 * http://creativecommons.org/licenses/by-sa/2.0/
 *
 * @author D. Jay Newman
 */
public class MapCell {
    // Occupancy
    public final int UNEXPLORED = -1;
    public final int INITIAL_VALUE = 128;
    public final int OCCUPIED = 2048;
    public final int FREE = 0;
```

```
// Surface textures
public final int SURFACE_UNKNOWN = 0;
public final int SURFACE_MIXED = 1;
public final int SURFACE_CARPET_SHORT = 10;
public final int SURFACE_CARPET_LONG = 11;
public final int SURFACE_HARD_SMOOTH = 20;

protected OccupancyMap map = null;
protected int surface = SURFACE_UNKNOWN;
protected int occupancy = UNEXPLORED;
protected int freedom = OCCUPIED;
protected int cost = 0;

/**
 *
 */
public MapCell() {
}

/**
 * getMap
 *
 * @return The OccupancyMap that may be in this cell
 */
public OccupancyMap getMap() {
        return map;
}

public void setMap(OccupancyMap om)  {
        map = om;
}

public boolean hasMap() {
        return map != null;
}

/**
 * setOccupancy()
 *
 * @param i The new occupancy level
 *
 * <p>
```

```
 * This is the probability of this cell somewhere between
 * FREE and OCCUPIED.
 *
 * @see MapCell#FREE
 * @see MapCell#OCCUPIED
 */
public void setOccupancy(int i) {
	occupancy = i;
}

public int getOccupancy() {
	return occupancy;
}

/**
 * setFreedom
 *
 * @param i The freedom level
 */
public void setFreedom(int i) {
	freedom = i;
}

public int getFreedom() {
	return freedom;
}

public void setCost(int i)  {
	cost = i;
}

public int getCost() {
	return cost;
}

public int getSurface() {
	return surface;
}

public void setSurface(int s) {
	surface = s;
}
```

```
}
```

AVRCam

The AVRCam is a nice camera available from http://www.jrobot.net/. It is similar in capabilities to the CMUCam.

These files are not currently used in Groucho, but they still work and compile.

ws.enerd.robots.avrcam.AVRCamConnection.java

```
/*
 * Created on May 3, 2005
 *
 * The licence is the Attribution-ShareAlike 2.0 License as stated at:
 * http://creativecommons.org/licenses/by-sa/2.0/
 */
package ws.enerd.robots.avrcam;

import java.io.*;
import java.util.*;

import gnu.io.*;

/**
 * AVRCamConnection
 *
 * <p>
 * This class represents a serial connection to an AVRCam.
 *
 * <p>
 * There are a couple of listeners and events used.
 *
 * <p>
 * Copyright 2005 D. Jay Newman
 *
 * <p>
 * The licence is the Attribution-ShareAlike 2.0 License as stated at:
 * http://creativecommons.org/licenses/by-sa/2.0/
 *
 * @author D. Jay Newman
 */
```

```
public class AVRCamConnection implements Runnable, AVRConstants {
   public static void main(String[] args) {
            RXTXCommDriver driver = new RXTXCommDriver();
            SerialPort port = (SerialPort) driver.getCommPort("/dev/tts/0",
                                CommPortIdentifier.PORT_SERIAL);

            try {
                     port.setSerialPortParams(115200, SerialPort.DATA-
BITS_8,
                                          SerialPort.STOPBITS_1,
SerialPort.PARITY_NONE);
                     AVRCamConnection connection = new
AVRCamConnection(port
                                          .getInputStream(),
port.getOutputStream());
                     connection.doPing();
            } catch (Exception ex) {
                     ex.printStackTrace();
            }
   }

   // The constants received from the camera
   private final byte EOL = (byte) 13; // end of the line
   private final byte[] ACK = { (byte) 'A', (byte) 'C', (byte) 'K', EOL}; //
Acknowledgement
   private final byte[] NCK = { (byte) 'N', (byte) 'C', (byte) 'K', EOL }; // No


                     // Acknoledgement

   // State Constants
   private final Object WAITING = new Object();

   private final Object ACK_RECEIVING = new Object();
   private final Object ACK_RECEIVED = new Object();

   private final Object NCK_RECEIVING = new Object();
   private final Object NCK_RECEIVED = new Object();

   // Commands sent to the camera
   private final byte[] PG = { (byte) 'P', (byte) 'G', EOL }; // Ping
```

```
private final byte[] GV = { (byte) 'G', (byte) 'V', EOL }; // Get Version
private final byte[] DF = { (byte) 'D', (byte) 'F', EOL }; // Dump a frame

// Command constants
private final Object CMD_PING = new Object();
private final Object CMD_VERSION = new Object();
private final Object CMD_FRAME = new Object();

private Object command = null;
private List listeners = new ArrayList();

// Instance variables
protected InputStream in = null;
protected OutputStream out = null;

public AVRCamConnection(InputStream in, OutputStream out) {
        this.in = in;
        this.out = out;
}

public void doPing() throws IOException {
        out.write(PG);
        waitForACK();
}

/**
 * waitForACK
 *
 * <p>
 * A Finite State Machine that reads in characters and
 * waits for either an ACK (acknowlegement) or a NCK (not-acknoledg-
ment)
 *
 * @return true for an ACK, false for a NCK
 * @throws IOException
 */
public boolean waitForACK() throws IOException {
        byte c;
        int i = 0;
        Object state = WAITING;
```

```
            while (true) {
                c = (byte)in.read();
                if (state == WAITING) {
                    if (c == ACK[0]) {
                        state = ACK_RECEIVING;
                    } else if (c == NCK[0]) {
                        state = NCK_RECEIVING;
                    }
                } else if (state == ACK_RECEIVING) {
                    if ((i >= ACK.length) || (c != ACK[i])) {
                        throw new IOException("BAD
ACK");
                    }

                    if (i == (ACK.length - 1)) {
                        state = ACK_RECEIVED;
                        break;
                    }
                } else if (state == NCK_RECEIVING) {
                    if ((i >= NCK.length) || (c != NCK[i])) {
                        throw new IOException("BAD
NCK");
                    }

                    if (i == (NCK.length - 1)) {
                        state = NCK_RECEIVED;
                        break;
                    }
                }
                i++;
            }

            return state == ACK_RECEIVED;
    }

    public void doDumpFrame() throws IOException {
            command = null;
            out.write(DF);
            if (!waitForACK()) {
                return;
            }
```

```
		for (int i = 0; i < HEIGHT; i++) {
			int b = in.read();
			if (b != 0x0b) {
				throw new IOException("Bad line : b = " + b);
			}

			int lineNum = readLineNumber();

			//System.out.println("Line num = " + lineNum);
			byte[] line = readScanLine();


			if (!readFrameEOL()) {
				throw new IOException("Bad end of scan
line");
			}


			AVRCamEvent event = new AVRCamEvent(this, true);
			event.setScanLine(new ScanLine(lineNum, line));
			fireListeners(event);
		}
	}

	/**
	 * readLineNumber
	 *
	 * <p>
	 * Read and return the line number (the lines in a frame may be
	 * unordered)
	 *
	 * @return
	 * @throws IOException
	 */
	protected int readLineNumber() throws IOException {
		while (in.available() < 1);
		return in.read();
	}

	/**
	 * readScanLine
	 *
```

```
 * <p>
 * Read a scan line
 *
 * @return
 * @throws IOException
 */
protected byte[] readScanLine() throws IOException {
        byte[] bytes = new byte[WIDTH];
        for (int i = 0; i < WIDTH; i++) {
                bytes[i] = (byte)in.read();
        }
        return bytes;
}

/**
 * readFrameEOL
 *
 * <p>
 * Read the end-of-line character
 *
 * @return true for an EOL and false for anything else
 * @throws IOException
 */
protected boolean readFrameEOL() throws IOException {
        int b = in.read();
        return b == 0x0F;
}

public boolean ping() {
        if (command == null) {
                command = CMD_PING;
                return true;
        }

        return false;
}

public boolean dumpFrame() {
        if (command == null) {
                command = CMD_FRAME;
                return true;
        }
```

```
            return false;
}

/**
 * @see java.lang.Runnable#run()
 */
public void run() {

            try {
                        while (true) {
                                    if (command == CMD_PING) {
                                                doPing();
                                    } else if (command == CMD_FRAME) {
                                                doDumpFrame();
                                    }

                                    command = null;
                        }
            } catch (Exception ex) {
                        ex.printStackTrace();
            }
}

public void addAVRCamListener(AVRCamListener listener) {
            listeners.add(listener);
}

private void fireListeners(AVRCamEvent event) {
            ListenerThread thread = new ListenerThread(listeners, event);
            thread.start();
}

private class ListenerThread extends Thread {
            private List listeners = null;
            private AVRCamEvent event = null;

            public ListenerThread(List listeners, AVRCamEvent event) {
                        this.listeners = listeners;
                        this.event = event;
            }
```

```
        /**
         * @see ws.enerd.util.StoppableThread#doRun()
         */
        public void run() {
                ListIterator iter = listeners.listIterator();

                while (iter.hasNext()) {
                        AVRCamListener listener =
(AVRCamListener) (iter.next());
                        listener.avrCamData(event);
                }
        }
    }
}
```

ws.enerd.robots.avrcam.AVRCamEvent.java

```
/*
 * Created on May 5, 2005
 *
 * The licence is the Attribution-ShareAlike 2.0 License as stated at:
 * http://creativecommons.org/licenses/by-sa/2.0/
 */
package ws.enerd.robots.avrcam;

import java.util.EventObject;

/**
 * AVRCamEvent
 *
 * <p>
 * Basically this is in case I need to do more with listeners
 * later.
 *
 * <p>
 * It will be sent to listeners at the end of each scan line.
 *
 * <p>
 * Copyright 2005 D. Jay Newman
 *
 * <p>
 * The licence is the Attribution-ShareAlike 2.0 License as stated at:
```

```
 * http://creativecommons.org/licenses/by-sa/2.0/
 *
 * @author D. Jay Newman
 */
public class AVRCamEvent extends EventObject {
    private ScanLine scanLine = null;
    private boolean success = true;

    public AVRCamEvent(Object src, boolean suc) {
            super(src);
            this.success = suc;
    }

    public void setScanLine(ScanLine sl) {
            scanLine = sl;
    }

    public ScanLine getScanLine() {
            return scanLine;
    }

    public void setSuccess(boolean b) {
            success = b;
    }

    public boolean getSuccess() {
            return success;
    }
}
```

ws.enerd.robots.avrcam.AVRCamFrameListener.java

```
/*
 * Created on May 8, 2005
 *
 * The licence is the Attribution-ShareAlike 2.0 License as stated at:
 * http://creativecommons.org/licenses/by-sa/2.0/
 */
package ws.enerd.robots.avrcam;

/**
 * AVRCamFrameListener
```

```
 *
 * <p>
 * Copyright 2005 D. Jay Newman
 *
 * <p>
 * The licence is the Attribution-ShareAlike 2.0 License as stated at:
 * http://creativecommons.org/licenses/by-sa/2.0/
 *
 * @author D. Jay Newman
 */
public interface AVRCamFrameListener {
    public void frameFinished(AVRCamImageEvent event);
}
```

ws.enerd.robots.avrcam.AVRCamImageEvent.java

```
/*
 * Created on May 8, 2005
 *
 * The licence is the Attribution-ShareAlike 2.0 License as stated at:
 * http://creativecommons.org/licenses/by-sa/2.0/
 */
package ws.enerd.robots.avrcam;

import java.util.EventObject;
import java.awt.image.*;

/**
 * AVRCamImageEvent
 *
 * <p>
 * Copyright 2005 D. Jay Newman
 *
 * <p>
 * The licence is the Attribution-ShareAlike 2.0 License as stated at:
 * http://creativecommons.org/licenses/by-sa/2.0/
 *
 * @author D. Jay Newman
 */
public class AVRCamImageEvent extends EventObject {
    private BufferedImage image;
```

```
    public AVRCamImageEvent(Object src, BufferedImage bi) {
            super(src);
            image = bi;
    }

    public BufferedImage getImage() {
            return image;
    }
}
```

ws.enerd.robots.avrcam.AVRCamImageReceiver.java

```
/*
 * Created on May 6, 2005
 *
 * The licence is the Attribution-ShareAlike 2.0 License as stated at:
 * http://creativecommons.org/licenses/by-sa/2.0/
 */
package ws.enerd.robots.avrcam;

import java.awt.image.*;
import java.util.*;

/**
 * AVRCamImageReceiver
 *
 * <p>
 * Copyright 2005 D. Jay Newman
 *
 * <p>
 * The licence is the Attribution-ShareAlike 2.0 License as stated at:
 * http://creativecommons.org/licenses/by-sa/2.0/
 *
 * @author D. Jay Newman
 */
public class AVRCamImageReceiver implements AVRCamListener,
AVRConstants, Runnable {

    private int linesReceived = 0;
    private BufferedImage image = null;
    private ArrayList listeners = new ArrayList();
    private int[] pixels = new int[IMAGE_HEIGHT * IMAGE_WIDTH];
```

```
    public AVRCamImageReceiver() {
    }

    public void startReceiving() {
            linesReceived = 0;
            image = new BufferedImage(IMAGE_WIDTH, IMAGE_HEIGHT,
BufferedImage.TYPE_INT_RGB);
    }

    /**
     * avrCamData
     *
     * <p>
     * This method recieves the pixels from the camera.
     *
     * <p>
     * Each byte is two pixels, each four bits. The even bytes contain GB and
the
     * odd bytes contain GR.
     *
     * <p>
     * Each byte[] is actually two pixel rows. The even columns the G is in
the
     * even row, and the B in the pixel below. The odd colums the G is in the
     * following row and the R in the even row.
     *
     * @see
ws.enerd.robots.avrcam.AVRCamListener#avrCamData(ws.enerd.robots.av
rcam.AVRCamEvent)
     */
    public void avrCamData(AVRCamEvent event) {
            // Currently the only type of event I'm looking for
            // is a Scan line received

            int b;

            ScanLine line = event.getScanLine();

            if (line != null) {
                    //System.out.println("Getting line: " +
line.getLineNum());
```

```
                    int n = line.getLineNum() * 2;

                    int even_row = n * IMAGE_WIDTH;
                    int odd_row = even_row + IMAGE_WIDTH;

                    byte[] bytes = line.getLine();

                    for (int i = 0; i < WIDTH; i++) {
                            b = 0x00FF & bytes[i];

                            //System.out.println("avrCamData: i = " + i +
"; b = " + b);

                            if ((i & 1) == 0) {
                                    // Even columns
                                    pixels[even_row] = ((b & 0x00F0) <<
8); // Green
                                    pixels[odd_row] = ((b & 0x000F) <<
4); // Blue
                            } else {
                                    // Odd colums
                                    pixels[even_row] = ((b & 0x000F) <<
16); // Red
                                    pixels[odd_row] = ((b & 0x00F0) <<
8); // Green
                            }

                            even_row++;
                            odd_row++;
                    }

                    linesReceived++;
                    if (linesReceived == HEIGHT) {
                            System.out.println("Finishing with image!");
                            // Process the pixels
                            processImage();

                            // Put pixel array into the image
                            image.setRGB(0, 0, IMAGE_WIDTH,
IMAGE_HEIGHT, pixels, 0, 1);
                            fireFrameListeners(new
```

```
AVRCamImageEvent(this, image));
                        }
                }
    }

    protected void processImage() {
    }

    public void addCamFrameListener(AVRCamFrameListener listener) {
                listeners.add(listener);
    }

    private void fireFrameListeners(AVRCamImageEvent event) {
                ListenerThread thread = new ListenerThread(listeners, event);
                thread.start();
    }

    private class ListenerThread extends Thread {
                private List listeners = null;
                private AVRCamImageEvent event = null;

                public ListenerThread(List listeners, AVRCamImageEvent
event) {
                        this.listeners = listeners;
                        this.event = event;
                }

                /**
                 * @see ws.enerd.util.StoppableThread#doRun()
                 */
                public void run() {
                        ListIterator iter = listeners.listIterator();

                        while (iter.hasNext()) {
                                AVRCamFrameListener listener =
(AVRCamFrameListener) (iter.next());
                                listener.frameFinished(event);
                        }
                }
    }

    /* (non-Javadoc)
```

```
     * @see java.lang.Runnable#run()
     */
    public void run() {
            while (true);
    }

}
```

ws.enerd.robots.avrcam.AVRCamListener.java

```
/*
 * Created on May 6, 2005
 *
 * The licence is the Attribution-ShareAlike 2.0 License as stated at:
 * http://creativecommons.org/licenses/by-sa/2.0/
 */
package ws.enerd.robots.avrcam;

/**
 * AVRCamListener
 *
 * <p>
 * Copyright 2005 D. Jay Newman
 *
 * <p>
 * The licence is the Attribution-ShareAlike 2.0 License as stated at:
 * http://creativecommons.org/licenses/by-sa/2.0/
 *
 * @author D. Jay Newman
 */
public interface AVRCamListener {
    public void avrCamData(AVRCamEvent event);
}
```

ws.enerd.robots.avrcam.AVRConstants.java

```
/*
 * Created on May 7, 2005
 *
 * The licence is the Attribution-ShareAlike 2.0 License as stated at:
 * http://creativecommons.org/licenses/by-sa/2.0/
 */
```

```
package ws.enerd.robots.avrcam;

/**
 * AVRConstants
 *
 * <p>
 * Copyright 2005 D. Jay Newman
 *
 * <p>
 * The licence is the Attribution-ShareAlike 2.0 License as stated at:
 * http://creativecommons.org/licenses/by-sa/2.0/
 *
 * @author D. Jay Newman
 */
public interface AVRConstants {
    public final int WIDTH = 176;
    public final int HEIGHT = 72;

    public final int IMAGE_WIDTH = 176;
    public final int IMAGE_HEIGHT = HEIGHT * 2;
}
```

ws.enerd.robots.avrcam.ScanLine.java

```
/*
 * Created on May 5, 2005
 *
 * The licence is the Attribution-ShareAlike 2.0 License as stated at:
 * http://creativecommons.org/licenses/by-sa/2.0/
 */
package ws.enerd.robots.avrcam;

/**
 * ScanLine
 *
 * <p>
 * Copyright 2005 D. Jay Newman
 *
 * <p>
 * Holds a scan line for the AVRCam
 *
 * <p>
```

```
 * The licence is the Attribution-ShareAlike 2.0 License as stated at:
 * http://creativecommons.org/licenses/by-sa/2.0/
 *
 * @author D. Jay Newman
 */
public class ScanLine {
   protected int lineNum = 0;
   protected byte[] line = null;

   public ScanLine(int lineNum, byte[] line) {
         this.lineNum = lineNum;
         this.line = line;
   }

   public int getLineNum() {
         return lineNum;
   }

   public byte[] getLine() {
         return line;
   }
}
```

ws.enerd.robots.avrcam.Test.java

```
/*
 * Created on May 8, 2005
 *
 * The licence is the Attribution-ShareAlike 2.0 License as stated at:
 * http://creativecommons.org/licenses/by-sa/2.0/
 */
package ws.enerd.robots.avrcam;

import gnu.io.*;


/**
 * Test
 *
 * <p>
 * Test the AVRCam interface
 *
```

```
 * <p>
 * Copyright 2005 D. Jay Newman
 *
 * <p>
 * The licence is the Attribution-ShareAlike 2.0 License as stated at:
 * http://creativecommons.org/licenses/by-sa/2.0/
 *
 * @author D. Jay Newman
 */
public class Test implements AVRCamFrameListener {

    public static void main(String[] args) {
            Test test = new Test("/dev/tts/0");
            test.dumpFrame();
    }

    private AVRCamConnection connection = null;
    private AVRCamImageReceiver receiver = null;

    private SerialPort port = null;
    private int i = 0;

    public Test(String portName)  {
            System.out.println("Test");

            RXTXCommDriver driver = new RXTXCommDriver();
            port = (SerialPort)driver.getCommPort(portName,
                            CommPortIdentifier.PORT_SERIAL);
            //port.setInputBufferSize(10000);

            try  {
                    port.setSerialPortParams(115200, SerialPort.DATA-
BITS_8,
                                    SerialPort.STOPBITS_1,
SerialPort.PARITY_NONE);
                    port.setFlowControlMode(SerialPort.FLOWCON-
TROL_NONE);
                    connection = new
AVRCamConnection(port.getInputStream(), port.getOutputStream());
            }
            catch (Exception ex)  {
                    ex.printStackTrace();
```

```
			}

			receiver = new AVRCamImageReceiver();
			new Thread(receiver).start();

			receiver.startReceiving();

			new Thread(connection).start();

			connection.addAVRCamListener(receiver);
			receiver.addCamFrameListener(this);
	}

	public void dumpFrame() {
			try {
					System.out.println("Time: " +
System.currentTimeMillis());
					receiver.startReceiving();
					connection.dumpFrame();
			}
			catch (Exception ex) {
					ex.printStackTrace();
			}
	}

	public void frameFinished(AVRCamImageEvent event) {
			System.out.println("Frame " + i++ + " finished");
			try {
					Thread.sleep(10);
					dumpFrame();
			}
			catch (Exception ex) {
					ex.printStackTrace();
			}
	}
}
```

Appendix

B

Resources

There are many websites that I've used to buy robotics equipment:

- BDMicro: http://www.bdmicro.com

  AVR microcontroller systems. Especially of interest are the MAVRIC-IIB microcontroller board and the RX50 motor drivers.
- Zagros Robotics: http://www.zagrosrobotics.com

  Specializes in medium-sized robotic bases and also sells sensors.
- Budget Robotics: http://www.budgetrobotics.com

  A general-purpose robotics shop.
- Acroname: http://www.acroname.com

  Another general-purpose robotics shop.
- Digikey: http://www.digikey.com

  Electronic components.
- Mouser: http://www.mouser.com

  Electronic components.
- Logic Supply: http://www.logicsupply.com

  Mini-ITX motherboards, systems, and accessories.
- JRobot: http://www.jrobot.net/

  This is the creator of the AVRCam.

Index

Note: Boldface numbers indicate illustrations.

About the Author

D. Jay Newman is a programmer, writer, and robot enthusiast. He has been interested in computers and robotics for as long as he can remember. This is his first book.